D0523172

Chemistry for Nonchemists

Principles and Applications for Environmental Practitioners

FRANK R. SPELLMAN

Government Institutes
An imprint of
The Scarecrow Press, Inc.
Lanham, Maryland • Toronto • Oxford
2006

 **Government Institutes**

Published in the United States of America
by Government Institutes, an imprint of The Scarecrow Press, Inc.
A wholly owned subsidiary of
The Rowman & Littlefield Publishing Group, Inc.
4501 Forbes Boulevard, Suite 200
Lanham, Maryland 20706
http://govinst.scarecrowpress.com

PO Box 317
Oxford
OX2 9RU, UK

British Library Cataloguing in Publication Information Available

Library of Congress Cataloging-in-Publication Data

Spellman, Frank R.
 Chemistry for nonchemists : principles and applications for environmental
practitioners / Frank R. Spellman.
    p.   cm.
 Includes bibliographical references and index.
 ISBN-13 978-0-86587-899-0 (pbk. : alk. paper)
 ISBN-10 0-86587-899-4 (pbk. : alk. paper)
 1. Chemistry.   2. Environmental chemistry.   I. Title.
 QD31.3.S64   2006
 540—dc22   2005020462

∞ ™ The paper used in this publication meets the minimum requirements
of American National Standard for Information Sciences—Permanence of
Paper for Printed Library Materials, ANSI/NISO Z39.48-1992.
Manufactured in the United States of America.

Contents

Preface

Chemistry for Nonchemists: Principles and Applications for Environmental Practitioners provides an overview and reviews the important aspects of chemistry in a way that you can easily understand even if you have not taken any formal chemistry courses. In this text, I will define the multidisciplinary *environmental practitioner* as anyone engaged in work directly related to environmental management, planning, impact assessment, environmental protection, or environmental compliance, including such activities as permitting, compliance auditing, regulatory review, research teaching, engineering, design, quality assurance, and implementation of environmental protection and control. Moreover, I include the occupational health and safety professional who designs, implements, and evaluates a comprehensive health and safety program that maintains and enhances health, improves safety, and increases productivity in my definition of *environmental practitioner*. Lastly, I loosely define (i.e., with some license) anyone with an active interest in environmental concerns as an *environmental practitioner*.

Along with basic chemistry principles, this book provides a clear, concise presentation of the consequences of chemistry's interaction with the environment. Although it is primarily designed for professionals (especially practicing environmental professionals) who need a quick review of chemical principles, applications, and basic calculations, it is also designed for a more general audience. Even if you are not at a lab bench, I will provide you, the nonchemist, with the jargon, concepts, and key concerns of chemistry and the chemical industry. The book is compiled in an accessible, user-friendly format, unique in that it explains scientific concepts with little mathematics and physical science.

You will be introduced to the basic facts and principles of chemistry and given problems and exercises to enhance the learning process. The book also contains a detailed overview of chemistry and its principles, chemical nomenclature, chemical reactions, and how they apply to regulatory compliance programs under the various environmental, health, and safety laws.

Although I developed this book for practicing environmental professionals who need a firm understanding of chemistry and its impact on environmental compliance programs, it is accessible and germane to any environmental interest—no matter the specialty. It is formatted for rapid learning and the text features clear, concise objectives, summaries, and chapter review questions.

I will begin with a review of the periodic table, chemical structure and bonding, and definitions of atoms, elements, molecules, and compounds. I will focus on the fundamentals of organic and inorganic chemistry, the chemistry of water, and chemical reactions as they apply to your environmental regulatory compliance programs.

Later, I will move to more advanced discussions of solvents and solutes, atmospheric reactions to chemicals, and toxicological chemistry. The book concludes with in-depth sessions on advanced sampling and analysis, the complex reactions of metals, and chemistry's role in risk assessment. Information is also provided to assist you in determining the compatibility of chemicals you might commonly find at work and at home and describes good laboratory practices.

This book is designed to serve the needs of students, teachers, consultants, and technical personnel in environmental compliance positions who must review chemistry basics and fundamentals. In order to maximize the usefulness of the material contained in the text, I have written and presented it in plain English in a simplified and concise format.

To assure correlation to modern practice and design, I have provided illustrative problems in terms of commonly used chemical parameters.

Each chapter ends with a review of key terms and concepts and a chapter review test to help evaluate mastery of the concepts presented. Before going on to the next chapter, you should take the review test, compare your answers to the key provided in the appendix, and review the pertinent information for any problems you missed. If you miss many items, review the whole chapter.

✔ **Note:** The checkmark symbol displayed in various locations throughout this text indicates or emphasizes an important point or points to study carefully.

Again, this text is accessible to those who have no experience with chemistry. If you work through the text systematically, you will acquire an understanding of and skill in chemistry principles—adding a crucial component to your professional knowledge.

CHAPTER 1

Introduction

The chemists are a strange class of mortals, impelled by an almost insane impulse to seek their pleasure among smoke and vapor, soot and flame, poisons and poverty; yet among all these evils I seem to live so sweetly that I may die if I would change places with the Persian King.

—Johann Joachim Becher

Topic in This Chapter

- Why Study Chemistry?

Welcome to chemistry, the science concerned with the study of matter. All matter on Earth consists of chemicals. The following is a comprehensive definition of chemistry:

chem•is•try *n., pl.* –tries. **1.** the science that systematically studies the composition, properties, and activity of organic and inorganic substances and various elementary forms of matter. **2.** chemical properties, reactions, phenomena, etc.: *the chemistry of carbon.* **3. a.** sympathetic understanding; rapport. **b.** sexual attraction. **4.** the constituent elements of something; the chemistry of love. [1560–1600; earlier *chymistry*].

The first part of this definition captures many of the essential ingredients of chemistry (although definitions 3 and 4 might make a more entertaining text).

So, how would you define chemistry? Are you tempted to answer that it is just a prerequisite for other subjects of study? Chemistry is that, but much more than that, for it explains our world around us.

Chemistry is the science of materials that make up the physical world. Chemistry is a systematic study. Chemistry is the study of the composition and properties of matter. Chemistry is the study of the reactivity of substances. Chemistry is the study of organic and inorganic substances. Chemistry is the study of connections between the everyday world and the molecular world. Chemistry is so complex that no one person could expect to master all aspects of such a vast field, so it has been found convenient to divide chemists into specialty areas. For example:

Organic chemists study compounds of carbon. Atoms of this element can form stable chains and rings, giving rise to very large numbers of natural and synthetic compounds.

Inorganic chemists are interested in all elements, but particularly in metals, and are often involved in the preparation of new catalysts.

Biochemists concern themselves with the chemistry of the living world.

Physical chemists study the structures of materials, and rates and energies of chemical reactions.

Theoretical chemists, with the use of mathematics and computational techniques, derive unifying concepts to explain chemical behavior.

Analytical chemists develop test procedures to determine the identity, composition, and purity of chemicals and materials. New analytical procedures often discover the presence of previously unknown compounds.

You are beginning the study or review of one of the most interesting subjects that you will ever come across. Of course, what you get out of studying chemistry will depend on what you put into it. Chemistry can intrigue and enlighten you or it can confound and frustrate you. It all depends upon the effort you are willing to put into your studies. To support that effort, this book helps sift through all the technical information and explains the value of understanding the basic concepts of chemistry that are relevant to everyday life. If you keep an open mind, follow each lesson step by step, and complete the chapter review tests, this book will change the way you view the world.

As mentioned in the preface, *Chemistry for Nonchemists* targets a specific audience and a more general audience:

- Professionals (especially in environmental sciences) who need a quick review of chemical principles, applications, and basic calculations
- Nonchemists who need an information source
- First-year chemistry students who need a study companion
- General readers who are not satisfied with their knowledge level in their specific area of expertise alone but prefer to add knowledge on the basic tenets of chemistry to their repertoire

Key features of the text include:

- For the nonchemist, familiarity with the jargon, concepts, and key concerns of environmental chemistry as it relates to the environmental profession
- Interesting and up-to-date applications, with numerous solved examples and easy-to-follow, step-by-step solutions in the text
- Review of problem-solving techniques
- Extensive lists of references and additional reading
- Easy-to-understand tables, figures, and diagrams
- Easy-to-understand language, with points of caution/interest (key and important points) to avoid misunderstanding or misapplication
- Common examples provided to allow the reader to understand the context of the information and its relevance to everyday life

- Explanations of concepts without mathematics and with little physical science
- For those entering the environmental profession, insights to suggest paths of inquiry in terms of career choices and goals

Why Study Chemistry?

Why should we care about chemistry? Isn't it enough to know that we don't want unnecessary chemicals in or on our food or harmful chemicals in our air, water, or soil?

Chemicals are everywhere in our environment. The vast majority of these chemicals are natural. The chemist often copies from nature to create new substances that are often superior to and cheaper than natural materials. It is human nature to make nature serve us. Without chemistry (and the other sciences), we are at nature's mercy. To control nature (to the extent possible), we must learn its laws, and then use them, as we must.

Environmental professionals must learn the "laws" of chemistry and its uses, but they must know more. Environmental professionals must know the ramifications of chemistry when it is out of control. Chemistry properly used can perform miracles. Out of control, chemicals and their effects can be, and have been, devastating. Many of Occupational Safety and Health Administration (OSHA) regulations dealing with chemical safety and emergency response procedures for chemical spills resulted because of catastrophic events involving chemicals. For example, OSHA's 1910.1200 standard, Hazard Communication, and its 1910.119 standard, Process Safety Management (PSM), originated with the horrific results of the Bhopal, India, incident that occurred just after midnight on December 3, 1984.

In the Bhopal incident, a U.S.-based Union Carbide Corporation pesticide plant accidentally released approximately 40 metric tons of methyl isocyanate (MIC) into the atmosphere. The initial incident was a catastrophe for Bhopal, with an estimated two thousand casualties, a hundred thousand injuries, and significant damage to livestock and crops. The long-term health effects from such an incident are difficult to evaluate; however, the International Medical Commission on Bhopal estimated that as of 1994 upwards of fifty thousand people remained partially or totally disabled.[1]

> ✔ **Important Point:** Methyl isocyanate is an organic chemical that is used in the production of pesticides. Commonly referred as MIC, it is extremely toxic to humans. Short-term exposure may cause death or adverse health effects that include pulmonary edema (respiratory inflammation), bronchitis, bronchial pneumonia, and reproductive effects.

The ultimate cause of the Bhopal incident remains in contention. However, as mentioned, one of its effects was the development and implementation of U.S. environmental legislation and regulations, including the creation of the U.S. Chemical Safety Board.

Although the Bhopal incident was a wake-up call for the U.S. Congress and for

OSHA and EPA and environmental practitioners, many environmental professionals felt that a similar disaster could not occur in the United States. Ironically, even as the experts opined that it couldn't happen here, another accidental chemical release did occur in the United States, in August 1985. This incident also involved a Union Carbide plant. The plant, located in Institute, West Virginia, released a cloud of methylene chloride and aldicarb oxide that affected four neighboring communities and led to the hospitalization of over a hundred people. In the wake of this and the Bhopal incident, the U.S. Congress passed the 1986 Emergency Planning and Community Right to Know Act, implemented by the EPA.

The impact of the Bhopal incident and its ramifications is best stated by the Union Carbide Corporation: "The legacy of those killed and injured [at Bhopal] is a chemical industry that adheres voluntarily to strict safety and environmental standards—working diligently to see that an incident of this nature never occurs again."[2] Hazardous chemical incidents are not the only area of concern for environmental professionals (and the rest of us). The fact is, almost every pollution problem we currently face has a chemical basis.

For example, during the 1970s, the media began reporting about a new environmental hazard: acid deposition. They described it in grim terms as "acid rain" and in even in grimmer terms as "death from the sky."

Then came reports pointing out that water pollutants can harm aquatic life, threaten human health, or result in the loss of recreational or aesthetic potential. Surface water pollutants come from industrial sources, nonpoint sources (e.g., runoff), municipal sources, background sources, and other/unknown sources. All of the eight chief water pollutants—biochemical oxygen demand, nutrients, suspended solids, pH, oil and grease, pathogenic microorganisms, toxic pollutants, and nontoxic pollutants—are, in one way or another, linked to chemistry.

In the early 1990s, a Global Assessment of Soil Degradation study was conducted for the United Nations Environment Programme. This study pointed out that in recent decades nearly 11% of Earth's fertile soil has been so eroded, chemically altered, or physically compacted as to damage its original biotic function (i.e., its ability to process nutrients into a form usable by plants). About 3% of soil has been degraded virtually to the point where it can no longer perform that function.[3]

Environmental Health and Safety (EH&S) practitioners focus primarily on maintaining the safety and good health of the environment and workers. However, they also have an obligation to protect the public and the environment from chemical-generated pollution problems such as the greenhouse effect, ozone depletion, toxic wastes, groundwater contamination, air pollution, and acid rain, to mention a few. To deal effectively with these kinds of chemical-based problems, EH&S professionals must have at least a rudimentary understanding of the basic concepts of chemistry.

The book focuses on chemistry as it relates to humans. It presents some concepts of environmental chemistry, and chemical safety is discussed. The book also illustrates how some information on chemistry is applied to real-world decision-making. For example, you will gain a very basic understanding of the fundamentals of environmental chemistry so you can comprehend key environmental data found in Material Safety Data Sheets (MSDSs), laboratory analytical reports, and the chemistry behind environmental regulations. You will also gain a better understanding of key environmental issues such as persistence of PCBs and DDT, the origins of acid rain, the dangers of lead, asbestos, and radon exposure, and the reasons that CFCs destroy our protective ozone layer. This basic "applications oriented" text also gives you more confidence in your ability to comprehend and even enjoy studying a daunting topic such as chemistry.

Additional Reading

Browning, J.G. *Union Carbide: Disaster at Bhopal*, Bhopal, 1993.
Hodapp, D. *Right to Know Laws*, http://www.crc.losrios.cc.ca.us/hodappd/15a/intro.htm, 1999.
Montague, P. "Carbide Officials Face Homicide Charges in Bhopal, India, Court," *Rachel Hazardous Waste News #58*, Environmental Research Foundation, January 4, 1988.
Senese, F. *What Is Chemistry?* http://antoine.frostburg.edu/chem/101/intro/faq/what-is-chemistry.shtml (accessed 2004).

Notes

1. The Implications of the Industrial Disaster in Bhopal India. Hearing Before the Subcommittee on Asian and Pacific Affairs of the Committee on Foreign Affairs. House of Representatives, 98th Congress, 2nd Sess., December 12, 1984.

2. *Bhopal.* Union Carbide Corporation. www.bhopal.com/index.htm (accessed December 2002).

3. World Resources Institute. *World Resources 1992–93* (New York: Oxford University Press, 1992), 206.

CHAPTER 2

Atoms and Elements

Just as several adjectives together describe an object (color of object, how tall or short, how bulky or thin, how light or heavy, etc.), several *properties*, or characteristics, must be used in combination to adequately describe a kind of matter. The form of matter, too, is distinguished from another form by its properties. However, simply saying that something is a colorless liquid isn't enough to identify it as, say, alcohol. A lot of liquids (e.g., pure water) are colorless, as are many solutions. More details are needed before you can zero in on the identity of a substance. Chemists will, therefore, determine several properties, both chemical and physical, in order to characterize a particular sample of matter—to distinguish one form of matter from another. The following discussion describes the differences between the two kinds of properties, chemical and physical.

Topics in This Chapter

- Defining Matter
- Atomic Theory
- Electron Configuration
- The Periodic Table of Elements
- Advanced Applications of the Periodic Table

Defining Matter

A thorough understanding of matter—how it consists of elements that are built from atoms—is critical for grasping chemistry. *Matter* is anything that occupies space and has weight (mass). Matter (or mass-energy) is neither created nor destroyed during chemical change.

✔ **Key Point:** Matter is measured by making use of its two properties. Anything that has the properties of having weight and taking up space *must* be matter.

As mentioned, along with the properties of having weight and taking space, matter has two other distinct properties, *chemical* and *physical properties*. These properties are actually used to describe *substances*, which are definite varieties of matter. Copper, gold, salt, sugar, and rust are all examples of substances. All of these substances are uniform in their makeup. However, if we pick up a common rock from our garden, we cannot call the rock a substance because it is a mixture of several different substances.

✔ **Key Point:** A *substance* is a definite variety of matter, all specimens of which have the same properties.

✔ **Important Point:** Under various environmental regulations *chemical substances* are defined differently from the definition provided above. For example, in 40 CFR Section 710.2, "chemical substance" means any organic or inorganic substance of a particular molecular identity, including any combination of such substances occurring in whole or in part as a result of a chemical reaction or occurring in nature, and any chemical element or uncombined radical; except that chemical substance does not include: (1) Any mixture, (2) Any pesticide when manufactured, processed, or distributed in commerce for use as a pesticide, (3) Tobacco or any tobacco product, but not including any derivative products, (4) Any source material, special nuclear material, or by-product material, (5) Any pistol, firearm, revolver, shells, and cartridges, and (6) Any food, food additive, drug, cosmetic, or device, when manufactured, processed, or distributed in commerce for use as a food, food additive, drug, cosmetic, or device.

PHYSICAL PROPERTIES

Substances have two kinds of physical properties: *intensive* and *extensive*. Intensive physical properties include those features that definitely distinguish one substance from another. Intensive physical properties do not depend on the amount of the substance. Some of the important intensive physical properties are color, taste, melting point, boiling point, density, luster, and hardness. It takes a combination of several intensive properties to identify a given substance. For example, a certain substance may have a particular color that is common to it but not necessarily unique to it. A white diamond is, as its name implies, white. However, is another gemstone that is white and faceted to look like a diamond really a diamond? Remember that a diamond is one of the hardest known substances. To determine if the white faceted gemstone is really a diamond we would also have to test its hardness and not rely on its appearance alone. Some of the important intensive physical properties are defined as follows:

Density: the mass per unit volume of a substance. Suppose you had a cube of lard and a large box of crackers, each having a mass of 400 grams. The density of the crackers would be much less than the density of the lard because the crackers occupy a much larger volume than the lard occupies.

The density of an object or substance can be calculated by using the following formula:

$$\text{Density} = \frac{\text{mass}}{\text{volume}}$$

The density of water may be measured in pounds per cubic foot (lb/ft^3) or pounds per gallon (lb/gal). However, the density of a dry substance, such as sand, lime, and soda ash, is usually expressed in pounds per cubic foot, as is the density of a gas, such as methane or carbon dioxide.

As shown in table 2.1, the density of a substance like water changes slightly as the temperature of the substance changes. This happens because substances usually increase in volume (size), as they become warmer. Because of this expansion with warming, the same weight is spread over a larger volume, so the density is lower when a substance is warm than when it is cold.

Specific Gravity: the weight of a substance compared to the weight of an equal volume of water. A substance having a specific gravity of 2.5 weighs two and a half times more than water. This relationship is easily seen when a cubic foot of water, which weighs 62.4 pounds, is compared to a cubic foot of aluminum, which weighs 178 pounds. Aluminum is 2.7 times as heavy as water.

It is not that difficult to find the specific gravity of a piece of metal. All we need do is to weigh the metal in air, then weigh it under water. Its loss of weight is the weight of an equal volume of water. To find the specific gravity, divide the weight of the metal by its loss of weight in water:

$$\text{Specific gravity} = \frac{\text{Weight of a substance}}{\text{Weight of equal volume of water}}$$

EXAMPLE 2.1

Problem: Suppose a piece of metal weighs 110 pounds in air and 74 pounds under water. What is the specific gravity?

Solution: Step 1: 110 lb subtract 74 lb = 36 lb loss of weight in water. Step 2:

$$\text{Specific gravity} = \frac{110 \text{ lb}}{36 \text{ lb}} = 3.1$$

✔ **Key Point:** In a calculation of specific gravity, it is *essential* that the densities be expressed in the same units.

The specific gravity of water is one (1), the standard reference to which all other substances (i.e., liquids and solids) are compared. Any object that has a specific gravity greater than one (1) will sink in water. Considering the total weight and volume of a ship, its specific gravity is less than one; therefore, it can float.

For the environmental practitioner, specific gravity has a number of applications. For example, consider that a liquid chemical has been accidentally spilled into one (or

Table 2.1.   Relative densities.

Temperature (°F)	Specific Weight (lb/ff³)	Density (slugs/ff³)
32	62.4	1.94
40	62.4	1.94
50	62.4	1.94
60	62.4	1.94
70	62.3	1.94
80	62.2	1.93
90	62.1	1.93
100	62.0	1.93
110	61.9	1.92
120	61.7	1.92
130	61.5	1.91
140	61.4	1.91
150	61.2	1.90
160	61.0	1.90
170	60.8	1.89
180	60.6	1.88
190	60.4	1.88
200	60.1	1.87
210	59.8	1.86

all) of the three environmental mediums—atmosphere, water, and/or soil. The EH& S professional will refer to that chemical's material safety data sheet (MSDS). If a chemical is spilled into a water body, the environmental practitioner will want to know, among other things, the chemical's specific gravity to determine if the chemical will sink or float. Obviously, this information is important because emergency response procedures for a contaminant that sinks might be different from those for one that floats.

Hardness: commonly defined as a substance's relative ability to resist scratching or indentation. Actual hardness testing involves measuring how far an "indenter" can be pressed into a given material under a known force. A substance will scratch or indent any other substance that is softer. Table 2.2 is used for comparing the hardness of mineral substances.

Table 2.2   Moh's hardness scale.

Hardness (H)	Mineral
1	Talc
2	Gypsum (fingernail: H = 2.5)
3	Calcite (penny: H = 3)
4	Fluorite
5	Apatite
6	Feldspar (glass plate: H = 5.5)
7	Quartz
8	Topaz

Environmental practitioners should also be familiar with another definition of hardness. In water treatment, for example, hardness is a characteristic of water, caused primarily by calcium and magnesium ions. Water hardness can cause many maintenance problems, especially with piping and process components where scale buildup can occur.

Odor: characteristic of many substances. Some have pleasant odors, like amyl acetate (fruity, banana, fragrant, sweet, earthy odor); some have pungent odors, like ammonia; some have disagreeable odors, like acetic acid (vinegar, sour).

For the environmental practitioner, odor can be an important parameter or first indication of potential trouble or hazards. For example, environmental practitioners are often involved with *confined space entry operations*. By configuration (i.e., limited access, enough room for worker entry, not normally occupied, and the possibility of hazardous atmosphere) confined spaces are inherently dangerous. If, before entry or during entry, personnel detect a very distinct rotten-egg odor, this should alert them to the presence of hydrogen sulfide, a very toxic gas at certain concentrations. Thus, precautions as indicated in the MSDS for hydrogen sulfide must be followed to protect entrants.

Caution: Hydrogen sulfide (H_2S) at low concentration does emanate a characteristic rotten egg odor. However, continued exposure (or exposures to higher levels) could result in numbing of the sense of smell to the point where its odor cannot be detected—giving those exposed a false sense of security—a very dangerous situation, obviously.

Color: another physical property of substances. Unless color blind, most people are familiar with the colors of various substances. Pure water is usually described as colorless. Water takes on color when foreign substances such as organic matter from soils, vegetation, minerals, and aquatic organism are present.

Extensive physical properties are such features as *mass, volume, length,* and *shape.* Extensive physical properties are dependent upon the amount of the substance.

CHEMICAL PROPERTIES

The nonchemist often has difficulty in distinguishing the physical versus the chemical properties of a substance. One test that can help is to ask the question: Are the properties of a substance determined without changing the identity of the substance? If we answer *yes,* then the substance is distinguished by its physical properties. If we answer *no,* then we can assume the substance is defined by its chemical properties. Simply, the *chemical properties* of a substance describe its ability to form new substances under given conditions.

To determine the chemical properties of certain substances, you can observe how the substance reacts in the presence of air, acid, water, a base, and other chemicals. You can also observe what happens when the substance is heated. If you observe a change from one substance to another, we know that a *chemical change*, or a *chemical reaction*, has taken place.

✔ **Key Point:** The chemical properties of a substance may be considered to be a listing of all the chemical reactions of a substance and the conditions under which the reactions occur.

Another example can be used to demonstrate the difference between physical and chemical change. When a carpenter cuts pieces of wood from a larger piece of wood to build a wooden cabinet, the wood takes on a new appearance. The value of the crafted wood is increased as a result of its new look. This kind of change, in which the substance remains the same, but only the appearance is different, is called a *physical change*. When this same wood is consumed in a fire, however, ashes result. This change of wood into ashes is called a *chemical change*. *In a chemical change a new substance is produced.* Wood has the property of being able to burn. Ashes cannot burn.

KINETIC THEORY OF MATTER

There are three states of matter—*solids, liquids, and gases*. All matter is made of *molecules*. Matter is held together by attractive forces, which prevent substances from coming apart. The molecules of a solid are packed more closely together and have little freedom of motion. In liquids, molecules move with more freedom and are able to flow. The molecules of gases have the greatest degree of freedom and their attractive forces are unable to hold them together.

The Kinetic Theory of Matter is a statement of how we believe atoms and molecules behave and how it relates to the ways we have to look at the things around us. Essentially, the theory states that all molecules are always moving. More specifically, the theory says:

• All matter is made of atoms, the smallest bit of each element. A particle of a gas could be an atom or a group of atoms.
• Atoms have an energy of motion that we feel as temperature. At higher temperatures, the molecules move faster.

✔ **Interesting Point:** *Kinetic* comes from the Greek word that means "motion." *Theory* comes from the Greek word for "idea."

Matter changes its state from one form to another. Here are some examples of how matter changes its state:

• Melting is the change of a solid into its liquid state.
• Freezing is the change of a liquid into its solid state.
• Condensation is the change of a gas into its liquid form.
• Evaporation is the change of a liquid into its gaseous state.

• Sublimation is the change of a gas into its solid state and vice versa (without becoming liquid).

✔ **Key Point:** Water freezing at 32°F is an example of a physical property of water.

Atomic Theory

To this point we have described how chemical change involves a complete transformation of one substance into another. Now that we have established a basic understanding of the chemical change of substances, we need to look at the structure of matter.

The ancient Greek philosophers believed that all the matter in the universe was composed of four elements: earth, fire, air, and water. Today we know of more than 100 different elements and scientists are attempting to create additional elements in their laboratories. An *element* is a substance from which no other material can be obtained (see figure 2.1).

Today, we know that *atoms* are the basic building blocks of all *matter*. Atoms equal the smallest particle of matter with constant properties (see figure 2.2).

Atoms are so small that it would take approximately *two thousand million* atoms side by side to equal one meter in length! (See figure 2.3.)

Atoms are so small that scientists were forced to devise special weights and measures:

Mass/weight: atomic units (au)
» 1 au $= 1.6604 \times 10^{-24}$ g

Length: Angstrom (Å)
» 1 Å $= 10^{-8}$ cm

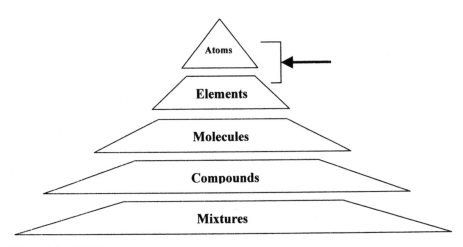

Figure 2.1. Matter.

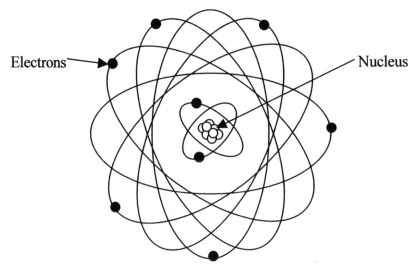

Figure 2.2. Atom.

Scientists used to believe that atoms were *indivisible*, but we now know that they are made up of many *subatomic particles*. Chemistry primarily deals with three subatomic particles: *protons, neutrons,* and *electrons.*

   Protons and *neutrons* are located in the nucleus (center) of the atom. *Electrons* are located in "orbitals" around the nucleus (see figure 2.4).

Particle	Charge	Mass
Proton (P)	+	1 atomic unit
Neutron (N)	no charge	1 atomic unit
Electron (e)	−	(none)

✔ **Key Point:** An atom contains these subatomic particles:
   • Protons (+), positively charged particles
   • Electrons (−), negatively charged particles
   • Neutrons (0), which have no charge
   The *stability* of a nucleus depends on the balance between attractive gravitational forces, repulsive electronic forces, and the ratio of protons to neutrons.

✔ **Key Point:** In a stable atom (or neutral atom), the number of electrons = the number of protons.

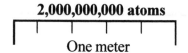

2,000,000,000 atoms

One meter

Figure 2.3.   2,000,000,000 atoms, side by side, equal one meter in length.

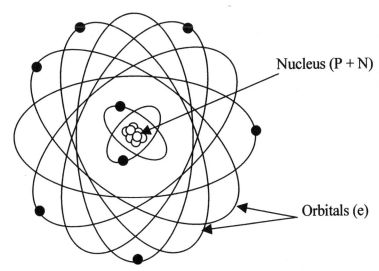

Nucleus (P + N)

Orbitals (e)

Figure 2.4. Orbitals.

ELEMENTS

As mentioned, scientists have identified more than a hundred different types of atoms, which they call *elements*. *An atom is the smallest unit of an element that still retains the properties of that element.*

Of the more than one hundred elements, only eighty-three are not naturally radioactive, and, of those, only fifty or so are common enough to human experience to be useful in this text. These elements, though, are going to stay the same for a long time; that is, they will outlast any political entity.

✔ **Key Point:** An *element* is a substance from which no other material can be obtained. Stated differently, the chemical elements are the simplest substances into which ordinary matter may be divided. All other materials have more complex structures and are formed by the combination of two or more of these elements.

Elements are substances; therefore they have physical properties that include density, hardness, odor, color, etc., and chemical properties that describe their ability to form new substances (i.e., a list of all the chemical reactions of that material under specific conditions).

Each element is represented by a *chemical symbol*. Chemical symbols are usually derived from the element's name (e.g., Al for aluminum). The chemical symbols for the elements known in antiquity are taken from their Latin names (e.g., Pb, from *plumbum*, for lead).

For every element there is one and only one uppercase letter (e.g., O for oxygen). There may or may not be a lower case letter with it (e.g., Cu for copper). When written in chemical equations, we represent the elements by the symbol alone with no charge attached.

EXAMPLES OF CHEMICAL SYMBOLS

Fe (iron)   P (phosphorus)
Al (aluminum)   Ag (silver)
Ca (calcium)   Cl (chlorine)
C (carbon)   Cu (copper)
N (nitrogen)   K (potassium)
Rn (radon)   He (helium)
H (hydrogen)   Si (silicon)
Cd (cadmium)   U (uranium)

✔ **Key Point:** Science recognizes only about a hundred elements (but well over a million compounds). Only 88 of the 100 + elements are present in detectable amounts on Earth, and many of these 88 are rare. Only ten elements make up approximately 99% (by mass) of Earth's crust, including the surface layer, the atmosphere, and bodies of water (see table 2.3).

As can be seen from table 2.3, the most abundant element on Earth is oxygen, which is found in the free state in the atmosphere as well as in combined form with other elements in numerous minerals and ores.

MATTER AND ATOMS

Molecules (see figure 2.5) consist of two or more atoms that have chemically combined

$$A + B \rightarrow C$$

Table 2.3   Elements making up 99% of Earth's crust, oceans, and atmosphere.

Element Number	Symbol	% of Composition	Atomic
Oxygen	O	49.5%	8
Silicon	Si	25.7%	14
Aluminum	Al	7.5%	13
Iron	Fe	4.7%	26
Calcium	Ca	3.4%	20
Sodium	Na	2.6%	11
Potassium	K	2.4%	19
Magnesium	Mg	1.9%	12
Hydrogen	H	1.9%	1
Titanium	i	0.58%	22

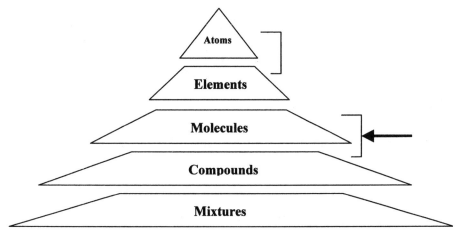

Figure 2.5.   Molecules.

When atoms (elements) chemically combine to form *compounds,* they *lose all their original properties* and create a new set of properties, unique to the compound (see figure 2.6). For example, sodium (Na) and chlorine (Cl) are poisonous, but they combine to form a compound called sodium chloride, which is ordinary table salt.

A molecule is the smallest unit of a compound that still retains the properties of that compound (see figure 2.7).

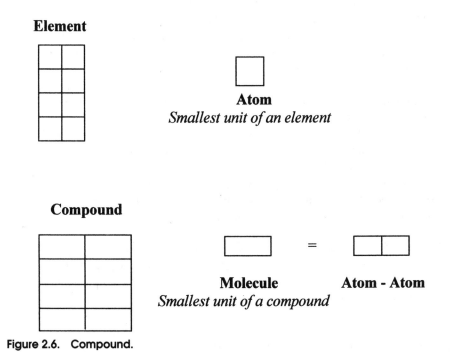

Figure 2.6.   Compound.

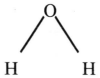

Figure 2.7. Water molecule.

Each individual compound will always contain *the same elements in the same proportions* by weight.

$$H_2O \text{ (water)} \neq H_2O_2 \text{ (hydrogen peroxide)}$$

Compounds are represented by *chemical formulas*. Chemical formulas consist of *chemical symbols* and subscripts to describe the relative number of atoms present in each compound. Common chemical formulas include:

- H_2O (water)
- NaCl (sodium chloride, table salt)
- HCl (hydrochloric acid)
- CCl_4 (carbon tetrachloride)
- CH_2Cl_2 (methylene chloride)

✔ **Key Point:** A chemical formula tells us how many atoms of each element are in the molecule of any substance. As mentioned, the chemical formula for water is H_2O. The "H" is the symbol for hydrogen. Hydrogen is a part of the water molecule. The "O" means that oxygen is also part of the water molecule. The "2" after the H means that two atoms of hydrogen are combined with one atom of oxygen in each water molecule.

Most naturally occurring matter consists of *mixtures* of elements and/or compounds (see figure 2.8). Mixtures are found in rocks, the ocean, vegetation, just about anything we find. Mixtures are combinations of elements and/or compounds held together by physical rather than chemical means. Mixtures are *physical combinations* and compounds are *chemical combinations*.

Mixtures have a wide variety of compositions. Mixtures can be separated into their ingredients by physical means (e.g., filtering, sorting, distillation, etc.). The *components of a mixture* retain their own properties.

✔ **Key Point:** Two classic examples of mixtures are concrete and salt water. Each appears to be a compound but is not, because physical forces (grinding, to break down cement, and filtering, to remove salt from water) can recover its component molecules just as they were before they were mixed.

Compounds are two or more elements that are "stuck" (bonded) together in definite proportions by a chemical reaction—for example, water (H_2O). Halite (table salt—

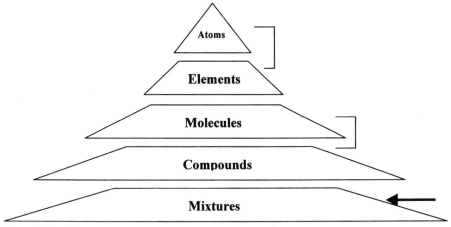

Figure 2.8.   Mixtures.

NaCl) is another example. Compounds always have uniform proportions, have unique properties that are different than their components, and cannot be separated by *physical* means.

Electron Configuration

Recall that an atom contains subatomic particles: protons ($+$), electrons ($-$), and neutrons (0). Free (unattached) uncharged atoms have the same number of electrons as protons to be electrically neutral. The protons are in the nucleus and do not change or vary except in some nuclear reactions. The electrons are in discrete pathways or *shells* (orbits) around the nucleus. There is a ranking or hierarchy of the shells, usually with the shells farther from the nucleus having a higher energy.

In an atom, the electrons seek out the orbits that are closest to the nucleus, because they are located at a lower energy level. The low-energy orbits fill first. The higher-energy levels fill with electrons only after the lower-energy levels are occupied. The lowest-energy orbit is labeled as the *K shell* (or shell 1), which is closest to the nucleus. (Note: The outer orbits, or shells, were historically named *K, L, M, N, O, P,* and *Q* for the chemists who found and calculated the existence of the first (inner) shells. The names of the first three chemists happened to begin with "K", then "L," then "M," respectively, so subsequent energy levels were continued up the alphabet. However, the numbers 1 through 7 have since been substituted for the letters.)

In the atomic diagram shown in figure 2.9, it can be seen that two electrons are needed to fill the K shell (shell 1), eight electrons to fill the L shell (shell 2), and, for light elements (atomic numbers 1–20), eight electrons will fill the M shell (shell 3).

For the following discussion, refer to the Electron Configuration Chart below. *Electron configuration* is the "shape" of the electrons around the atom, that is, which energy level (shell) and what kind of orbital it is in. The electron configuration is

K	L	M	N	O	P	Q	
1	2	3	4	5	6	7	
s	*sp*	*spd*	*spdf*	*spdf*	*spd*	*sp*	
2	8	8	2				20
		10	6	2			38
			10	6	2		56
			14	10	6	2	88
				14	10	6	
___	___	___	___	___	___	___	
2	8	18	32	32	18	8	TOTALS

Figure 2.9. Electron configuration chart.

written out with the first (large) number as the shell number. The letter indicates the orbital type (either *s*, *p*, *d*, or *f*). (The letters come from early twentieth-century terminology for the lines in atomic spectra.) The smaller superscript number is the number of electrons in that orbital.

To use this scheme, you first must know the orbitals. An *s* orbital only has 2 electrons. A *p* orbital has 6 electrons. A *d* orbital has 10 electrons. An *f* orbital has 14 electrons. We can tell what type of orbital it is by the number on the chart. The only exception to that is that "8" on the chart is "2" plus "6," that is, an *s* and a *p* orbital. The chart reads left to right and then down to the next line, just as in English writing. Any element with over 20 electrons in the electrical neutral unattached atom will have all the electrons in the first row on the chart.

The totals on the right indicate using whole rows. If an element has an atomic number over 38, take all the first two rows and whatever more from the third row. For example, iodine is number 53 on the periodic table of elements (discussed later). For its electron configuration we would use all the electrons in the first two rows and fifteen more electrons: 1*s*2 2*s*2 2*p*6 3*p*6 4*s*2 3*d*10 4*p*6 5*s*2 from the first two rows and 4*d*10 5*p*5 from the third row. We can add up the totals for each shell at the bottom. Full shells would give us the totals on the bottom.

✔ **Key Point:** Electron configuration is the "shape" of the electrons around an atom, that is, which energy level (shell) and what kind of orbital it is in.

The Periodic Table of Elements

The periodic table of elements is an arrangement of the elements into rows and columns in which those elements with similar properties occur in the same column. Chemists use this table to correlate the chemical properties of known elements and predict the properties of new ones.

✔ **Key Point:** The periodic table of elements arranges the elements on the basis of electronic distribution, to show a large amount of information and organization.

The periodic table of elements organizes information to provide the following:

- Atomic number (the number of electrons revolving about the nucleus of the atom)
- Isotopes (atoms of an element with different atomic weights)
- Atomic weight and molecular weight
- Groups (vertical columns) and periods (horizontal columns)
- Locating important elements

The periodic table of elements provides information about the element as shown in figure 2.10. Remember that each element is represented by a chemical symbol

Fe (iron)   P (phosphorus)
Al (aluminum)   Ag (silver)
Ca (calcium)   Cl (chlorine)
C (carbon)   H (hydrogen)

Atomic number is the *number of protons* in the nucleus of an atom (also the number of electrons surrounding the nucleus of an atom). Remember that protons are positively charged subatomic particles found in the nucleus of the atom.

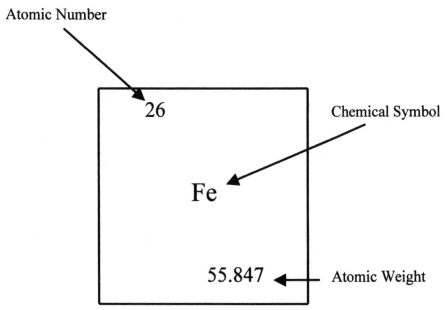

Figure 2.10. The element iron.

✔ **Key Point:** The number of protons in the nucleus determines the atomic number. Change the number of protons and you change the element, the atomic number, and the atomic mass.

As mentioned, the atomic number can also indicate the *number of electrons* if the *charge* of the atom is known.

Element	Atomic No.	Charge	No. of Electrons
O	8	0	8
O^{-2}	8	-2	10
Na	11	0	11
Na^+	11	$+1$	10

The atomic number (again, the number of protons) is a *unique identification number* for each element. It *cannot change* without changing the identity of the element.

$$C = 6 \quad O = 8 \quad Cl = 17$$

The *number of neutrons* can change without changing the identity of the element or its chemical and physical properties. Atoms with the same number of protons but different numbers of neutrons are called *isotopes*.

Isotope	#P	#N
^1H (hydrogen)	1	0
^2H (deuterium)	1	1
^3H (tritium)	1	2

If the number of neutrons changes, the *atomic weight* of the atom changes.

Isotope	#P	#N	Weight
^1H (hydrogen)	1	0	1 au
^2H (deuterium)	1	1	2 au
^3H (tritium)	1	2	3 au

Atomic weight is the relative weight of an average atom of an element, based on C^{12} being exactly 12 atomic units. Chemists generally *round* atomic weights to the nearest whole number.

$$H = 1 \quad O = 16 \quad Fe = 56$$

Chemists add up atomic weights to determine the *molecular weight* of a compound. For example, the molecular weight of water (H_2O) equals the weight of 2 hydrogen atoms and 1 oxygen atom.

$$\frac{1}{2(1.008)} + \frac{16}{1(15.999)} \cong 18$$

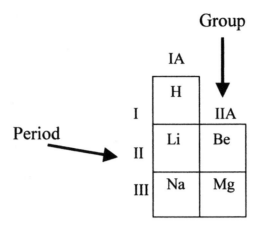

Figure 2.11.   Period and group.

Molecular weights are an important indication of the general size (and therefore complexity) of a compound. Important in the context of environmental science, in general (but not always), the higher the molecular weight, the greater the likelihood that the compound may persist in the environment.

Periods, the *rows* (horizontal lines) of the periodic table, are organized by increasing atomic number (see figure 2.11). *Groups* are the *columns* of the periodic table, which contain elements with similar chemical properties. Within each *period* (row), the *size* of the nuclei increases going from left to right because the *atomic number* increases (see figure 2.12).

Within each *group* (column), the *size* of the atoms increases going from top to bottom because the *number of electron shells* increases (see figure 2.13).

Within each *group* (column), elements have *similar chemical reactivity* because they have a *similar number of outer shell electrons* (see figure 2.14).

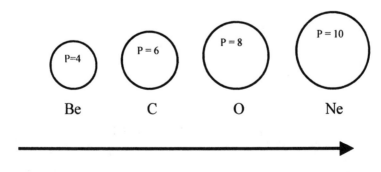

Increasing Nuclear Radius

Figure 2.12.   Nuclei increase with an increase in the atomic number.

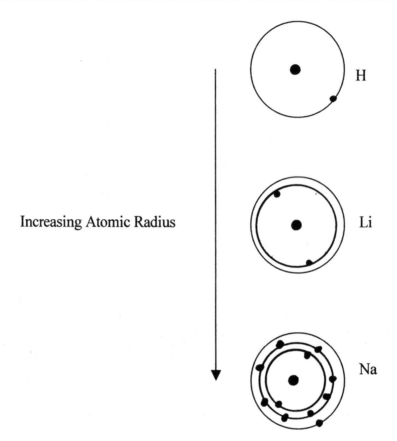

Increasing Atomic Radius

Figure 2.13. Atomic radius increase.

The *group numbers* (i.e., Roman numerals) above each column indicate the number of outer shell electrons in each group:

Group V = 5 outer shell electrons
Group II = 2 outer shell electrons
Group VII = 7 outer shell electrons

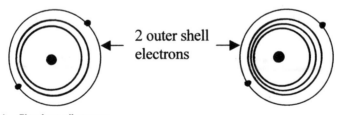

2 outer shell electrons

Figure 2.14. Electron diagram.

Only "outer shell" electrons are involved in chemical change. The nucleus and inner shell electrons are not altered in any way during ordinary chemical reactions (see figure 2.15).

Using the periodic table of elements, you should also be able to locate the following:

- The alkali metals (Group I)
- The halogens (Group VII)
- The noble gases (Group VIII)
- Metals, nonmetals, and metalloids
- The rare earths and radioactive elements

THE ALKALI METALS

- Lithium (Li)
- Sodium (Na)
- Potassium (Na)
- Rubidium (Rb)
- Cesium (Cs)

THE HALOGENS

- Fluorine (F)
- Chlorine (Cl)
- Bromine (Br)
- Iodine (I)

THE NOBLE GASES

- Helium (He)
- Neon (Ne)
- Argon (Ar)
- Krypton (Kr)
- Xenon (Xe)
- Radon (Rn)

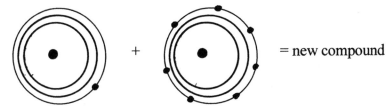

Figure 2.15. Ordinary chemical reaction.

METALS, NONMETALS, AND METALLOIDS

Typically, a diagonal line across the right-hand side of the periodic table separates the *metals* from the *nonmetals*. The *metalloids* are boron (B), silicon (Si), germanium (Ge), arsenic (As), antimony (Sb), and tellurium (Te).

THE RARE EARTHS AND RADIOACTIVE ELEMENTS

These elements are located in separate rows at the bottom of the periodic table of elements. The rare earths or *lanthanides* all resemble lanthanide in their chemical reactivity. The *actinides* resemble actinide reactivity, and all are radioactive.

✔ Key Point: The periodic table, because it is a systematic arrangement of the elements with similar chemical reactivities, allows chemists to predict trends in reactivity.

Advanced Applications of the Periodic Table

To this point we have presented a basic introduction to the periodic table—the recipe of the universe. In this section, we apply what we have learned about the periodic table to advanced applications.

Remember that the periodic table provides listing of each element and its chemical symbol (i.e., Fe [iron], Al [aluminum], C [carbon], etc.). Along with the element's name and chemical symbol, the atomic number and atomic weight are also listed (e.g., atomic number for Fe is 26; the atomic weight for Fe is 55.847).

The *atomic number* is the number of protons in the nucleus of an atom and is a *unique identification number* for each element. It *cannot* change without changing the identity of the element.

$$C = 6 \quad O = 8 \quad Cl = 17$$

The *number of neutrons* in the nucleus *can* change without changing the element's identity. As mentioned, isotopes are atoms with the same number of protons but different numbers of neutrons.

ELECTRON SHELLS

In the periodic table of elements, note that shell 1 is very small and can only hold up to 2 electrons. Moreover, notice that period 1 (row 1) only contains 2 elements: H with 1 electron and He with 2 electrons. Period 1 (row 1) contains only 2 elements because shell 1 can only hold 2 electrons.

Shell 2 is larger and can hold up to 8 electrons. Period 2 (row 2) contains 8 elements: Li, Be, B, C, N, O, F, and Ne. Period 2 (row 2) contains only 8 elements because shell 2 can only hold 8 electrons. This pattern continues with each subsequent shell as follows:

- Shell 3 can hold up to 8 electrons.
- Shell 4 can hold up to 18 electrons.
- Shell 5 can hold up to 18 electrons.
- Shell 6 can hold up to 32 electrons.
- Shell 7 can hold up to 32 electrons.

In general, each shell must be *filled* before any electrons go into the next shell.

ELECTRON ORBITALS

Remember that there are up to four different types of *electron orbitals* (subshells) within each shell, which, as noted above, chemists refer to as *s, p, d* and *f*.

✔ **Key Point:** Each orbital can hold 2 electrons.

The *s* orbital is spherical. There are 3 different *p* orbitals, which are shaped like elongated dumbbells and can hold a total of 6 electrons (i.e., 2 electrons each). The five *d* orbitals and seven *f* orbitals are more complex in shape and can hold 10 and 14 electrons, respectively.

Let's review the pattern for filling electron shells as described earlier:

Shell 1 can hold 2 electrons: One *s* orbital = 2 electrons.

Shells 2 and 3 can hold 8 electrons each: One *s* and three *p* orbitals = 8 electrons (i.e., 4 orbitals, each holding 2 electrons).

Shells 4 and 5 can hold 18 electrons each: One *s*, three *p*, and five *d* orbitals = 18 electrons (i.e., 9 orbitals, each holding 2 electrons).

Shells 6 and 7 can hold 32 electrons each: One *s*, three p, five *d*, and seven *f* orbitals = 32 electrons (i.e., 16 orbitals, each holding 2 electrons).

THE ORGANIZATION OF THE PERIODIC TABLE

Atoms "prefer" to have filled outer shells. Therefore the noble gases (Group VIIIA) are chemically unreactive because they have filled outer shells (a highly preferred state).

In contrast, the *halogens* (Group VIIA) are very chemically reactive because they lack only one electron to fill their outer shells. This is why they are highly *electronegative* atoms.

The *alkali metals* (Group IA) are also very chemically reactive because they have only one outer shell electron.

Within each group (column), elements have *similar chemical reactivity* because they have a similar number of outer shell electrons *in the same type of orbital.* The group numbers (i.e., Roman numerals) above each column indicate the *number of outer shell electrons* in each group.

Group V = 5 outer shell electrons
Group II = 2 outer shell electrons
Group VII = 7 outer shell electrons

The *letters* associated with each group number indicate the type of orbital.

A = *s* and/or *p* orbitals
B = *d* and/or *f* orbitals
Group II*A* = *2* outer shell electrons in *s* orbitals
Group III*B* = *3* outer shell electrons in *d* plus *f* orbital electrons (the lanthanides and actinides)
Group VIII*B* = *8* outer shell electrons in *d* orbitals

✔ **Key Point:** The periodic table is a systematic arrangement of the elements, which groups elements with similar chemical reactivities and allows chemists to predict trends in reactivity.

Additional Reading

Asimov, I., ill. D. F. Back. *Atom: Journey across the Subatomic Cosmos.* New York: Dutton/ Plume, 1992.
Levi, Rimor, trans. R. Rosenthal. *The Periodic Table (American).* New York: Knopf Publishing Group, 1986.
Mebane, R.C., and T. R. Rybolt. *Adventures with Atoms and Molecules: Chemistry Experiments for People.* Vol. 1. Berkeley Heights, New Jersey: Enslow Publishers, 1998.
Nardo, D. *Atoms.* New York: Gale Group, 2001.
Stwertka, A. *Guide to the Elements.* New York: Oxford University Press, 2002.

Summary

You should now be able to associate:

- *Atomic number* with number of protons in nucleus and identity of element
- *Neutrons* with isotopes (two atoms of the same element are isotopes if they differ in the number of neutrons in their nuclei)
- *Mass number* with number of protons plus number of neutrons
- *Atomic weight* with average of mass numbers for all isotopes of an element as they occur on Earth

New Word Review

Alkali metals—the elements in Group I. They are very active metals that react with water to form strong bases.

Alkaline earth metals—the elements in Group II in the periodic table. Their oxides crumble and feel like dry earth.

Atoms—the units of an element that make up molecules.

Atomic nucleus—the central part of an atom that contains the protons and neutrons.

Atomic number—the number of protons in the nucleus.

Atomic weight—the net weight of all protons and neutrons in the nucleus.

Electron—a particle that is part of all atoms. It has a negative electrical charge and almost no weight.

Elements—substances whose atoms have the same atomic number.

Energy level—the orbit, or shell, in which the electron is located.

Gas—matter that does not occupy a definite volume, and is shapeless.

Group—a family (vertical list) of elements in the periodic table.

Isotopes—atoms with the same atomic number, but a different atomic weight.

Kinetic molecular theory—the idea that molecules are in constant motion.

Liquid—matter that occupies a definite volume but is shapeless.

Mass number—the sum of all protons and neutrons in the nucleus.

Melting point—the temperature at which a solid changes into a liquid.

Metal—an element whose atoms release valence electrons when they react.

Metalloids—elements that resemble both metals and nonmetals. They have properties found in metals and in nonmetals.

Molecule—the smallest particle of a substance that can exist with its own identity.

Neutron—a particle that is a part of the atom. It has no electrical charge and has about the same weight as the proton.

Nonmetal—an element whose atoms gain valence electrons when they react.

Period—the horizontal listing of elements in the periodic table.

Periodic table—the arrangement of all the elements by atomic numbers. Elements listed in vertical columns are in the same family because they share similar properties.

Proton—a particle that is a part of all atoms. It has a positive electrical charge and weighs almost 2,000 times more than an electron.

Solid—matter that occupies a definite space (volume) and has a definite shape.

Chapter Review Questions

2.1. A material from which no new material can be obtained is a(n) _____.

2.2. The _____ has a positive charge.

2.3. The _____ has a negative charge.

2.4. The _____ has no mass.

2.5. When atoms combine they form _____.

2.6. The _____ has no electrical charge.
2.7. The _____ and _____ have equal masses.
2.8. The number of _____ determines what the element will be.
2.9. The sulfur atom, with 16 protons and 16 neutrons, has an atomic mass of: _____.
2.10. An atom with 12 protons and 13 neutrons must have _____ electrons.
2.11. As the number of neutrons in the nucleus increases, the: _____.
2.12. The maximum number of electrons that can fill the L shell is: _____.
2.13. Using the periodic table of elements, what is the molecular weight of CO_2? _____.
2.14. Fill in the following table by referring to the periodic table of elements.

Element	No. of Protons	No. of Electrons	Charge
Ag			
K^+			
Br^-			
Ca^{+2}			

2.15. Which of the following are physical changes and which are chemical changes?
  a. Snow melting _____
  b. Milk souring _____
  c. Sodium bicarbonate neutralizing an "acid" stomach _____
  d. Gas bubbling out of ginger ale _____
  e. The disappearance of sugar when it dissolves in water _____
2.16. Some examples of physical properties are _____.
  a. mass and weight
  b. volume and density
  c. color, shape, hardness, and texture
  d. all of the above
2.17. Chemical properties: _____.
  a. are changes in phase
  b. describe how a substance changes in other substances
  c. are changes in volume or shape
  d. are changes in mass of weight
2.18. An example of a chemical change is: _____.
  a. burning paper
  b. dissolving sugar in tea
  c. mixing alcohol
  d. melting ice
2.19. An example of a chemical property is: _____.
  a. color
  b. the ability to rust
  c. density
  d. mass

2.20. The characteristic of matter that is not a physical property is: _____.
   a. hardness
   b. flammability
   c. shape
   d. texture

2.21. A new substance is formed as a result of: _____.
   a. physical change
   b. chemical change
   c. physical property
   d. chemical property

CHAPTER 3

Chemical Bonds

The joining of two or more atoms forms chemical compounds. A stable compound occurs when the total energy of the combination has lower energy than the separated atoms. The bound state implies a net attractive force between the atoms . . . a chemical bond.

Topics in This Chapter

- How Atoms Are Linked Together
- About Chemical Bonds
- Chemical Formulas

Now that you understand what atoms are and how they look and have a basic understanding of the periodic table, I will begin this chapter on chemical bonds with a brief review of key points to set the stage for the material presented.

- *Matter* is anything that takes up space and has mass.
- An *element* is a substance that cannot be broken down by ordinary means, the material making up matter.
- Small units of matter are called *atoms*. *Protons* ($+$), *neutrons* (0), and *electrons* ($-$) are the subunits of atoms.
- *Atomic number* is the total number of protons in an atom.
- *Atomic mass* is the total number of neutrons and protons in an atom.
- *Isotopes* are different atomic forms caused by the variance of the number of electrons.
- *Energy levels*: All electrons have the same mass and charge; they simply differ in the amounts of potential energy they possess. Electrons closer to the nucleus contain less potential energy.
- *The periodic table* is a chart made up of 18 columns and 7 rows. The elements on the table are organized by atomic number. The periodicity of the elements defines unique properties, so elements of a given column in the table have similar properties.

As mentioned earlier, atoms of elements can combine to form molecules of compounds. Experimentation has shown that only certain combinations of atoms will bond together. It has also been found through experiments that the number of atoms of each element in a mole is very definite. The best example of this is a water molecule, which is formed of two hydrogen atoms and one oxygen atom—no other combination of hydrogen and oxygen makes water. In this chapter, I will provide a brief discussion of why only certain molecules occur, the different types of bonds, and chemical formulas.

How Atoms Are Linked Together

The tendency of elements to link together to form compounds through a shift of electronic structure is known as *valence*. This linking process is accomplished through *valence electrons*. These electrons occupy the last energy level of an atom. It is here where atoms come in contact with each other. It stands to reason that chemical bonds will occur here in any chemical reaction. The maximum number of valence electrons any atom can contain is 8. Any number less than 8 will allow that atom to act as a donor or recipient of electrons to become stable. Atoms that give electrons will become positive ions and have a positive (+) charge, while atoms that receive electrons will become negative ions with a negative (−) charge.

About Chemical Bonds

Atoms are linked or joined by *chemical bonds*. Only electrons are involved in the formation of chemical bonds between atoms. Only the *outermost* electrons (i.e., the *outer-shell electrons*) are typically involved in bonding. Each bond consists of two electrons, one from each atom in the bond.

There are different "types" of chemical bonds—depending on the type of atoms that are bonded together—covalent and ionic bonds.

A *covalent bond* results from the sharing of a pair of electrons between two covalent bonds:

Nonpolar: Here the valence electrons are shared equally, thus eliminating a positive and negative end on the molecule. An example of a nonpolar bond is $CHCl_4$. It is nonpolar because although the 4 C-Cl bonds are all polar, the symmetry of their arrangement around the central C atom makes the overall molecule nonpolar.

Polar covalent bond: Here the valence electrons are shared unequally, causing the molecule to develop a positive end (where the electrons spend less time) and a negative end (where the electrons spend more time). This has to do with the electronegativity of the atom. The more electronegative the atom, the more it will hold onto the electrons. Oxygen is very electronegative. Hydrogen is not. If the difference in electronegativity between two atoms is sufficiently large, the shared electron pair will spend all of its time on the more electronegative atom, resulting in ionic bonding rather than covalent bonding. An example of a polar covalent bond is $CHCL_3$, because of the polar C-Cl bond (Cl is an electronegative atom).

✔ **Key Point:** A polar covalent bond has a positively charged end and a negatively charged end.

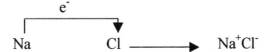

Figure 3.1. Ionic bond.

In covalent bonding, the sharing of one pair of electrons is called a *single bond*; the sharing of two pairs of electrons is called a *double bond (e.g.,* carbon dioxide) and the sharing of three pairs of electrons is called a *triple bond* (e.g., acetylene). Double bonds are more reactive than single bonds and compounds containing double bonds are somewhat more volatile than corresponding single-bonded molecules. Triple bonds are even more reactive than double bonds, and volatility in triple-bonded compounds is still greater.

An *ionic bond* results from the transfer of electrons from one atom to another.

In an ionic bond, one atom donates one or more of its outermost electrons to another atom or atoms (see figure 3.1). The atom that gains the electrons becomes a negative ion, or *anion*.

✔ **Key Point:** Remember the mnemonic "*a* negative *ion*."

The atom that loses the electron becomes a positive ion, or *cation*.

✔ **Key Point:** Covalent bonding is a process similar to ionic bonding, but the electrons are shared rather than transferred.

ELECTRONEGATIVITY AND POLARITY

Electronegativity is a measure of the ability of an atom to attract its outermost electrons. The higher the electronegativity, the greater the atom's ability to attract other electrons. Common electronegative atoms are shown in figure 3.2.

Highly electronegative atoms tend to form *ionic,* or *polar, covalent* bonds. In polar covalent bonds, the more electronegative atom "keeps" the electron pair more (i.e., takes a larger share of the electron density). In nonpolar covalent bonds (e.g., C-C or C-H), the electrons are shared equality.

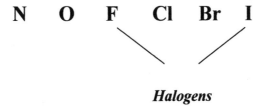

Figure 3.2. Common electronegative atoms.

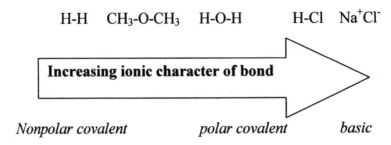

Figure 3.3. Electronegativity and polarity.

✔ **Key point:** Think of chemical bonding as a continuum from nonpolar covalent bonds to ionic bonds (see figure 3.3). Bond polarity depends on the *electronegativity* of the atoms involved, but overall molecular polarity depends on *symmetry* (see figure 3.4).

BOND STRENGTHS AND BOND ANGLES

Chemists measure the *strength* of a bond by determining how much energy is needed to break the bond (see figure 3.5). Other examples of carbon (C) bond strengths are shown below.

Type of Bond	Strength (kcal/mol)
C-N	70
C-Cl	79
C-O	84
C-H	99

In addition to measuring bond strengths, chemists can also measure *bond angles* between atoms (see figure 3.6). Bond angles depend on the type of atoms bonded to-

Figure 3.4. Bond polarity.

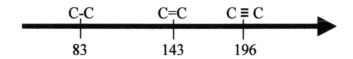

C-C   C=C   C ≡ C

83   143   196

Increasing Bond Strength

(kcal/mole)

Figure 3.5.   Bond strength.

180°

Cl----Be----Cl

F
|
120° (B

F   F

H

H ▬ C) 109.5°

H   H

Figure 3.6.   Bond angles.

gether. Chemists predict bond angles and other 3D structures using computer programs (see figure 3.7).

STRUCTURAL FORMULAS

The overall 3D shape of a molecule is represented by its *structural formulas,* which shows how the atoms are attached (see figure 3.8). In chemical shorthand, the Cs and Hs are often not shown (see figure 3.9). *Double bonds* and *triple bonds* are shown in figure 3.10.

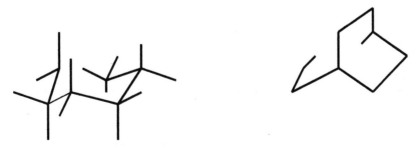

Figure 3.7.   Structure and bonding.

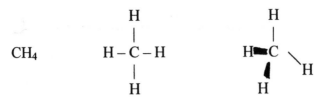

Figure 3.8.   Sample structural formulas.

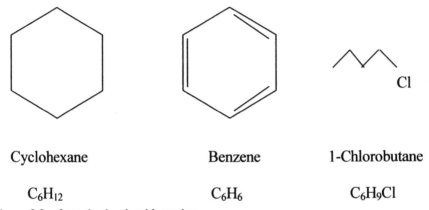

Cyclohexane   Benzene   1-Chlorobutane

C_6H_{12} C_6H_6 C_6H_9Cl

Figure 3.9.   Sample structural formulas.

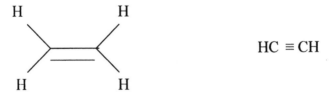

Double Bond   Triple Bond

Figure 3.10.   Sample structural formulas.

ISOMERS

Isomers are defined as one of several organic compounds with the same chemical formula but different structural formulas and therefore different properties (see figure 3.11).

✔ **Key Point:** Structural formulas give us the same information as the chemical formulas but they also tell us how the atoms are bonded together.

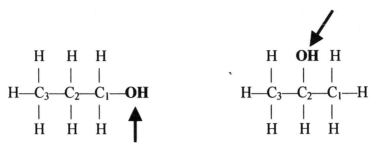

Figure 3.11. Isomers.

As mentioned, isomers are different compounds that have different properties, although these differences are often slight (see figure 3.12).

✔ **Key Point:** Structure makes the difference!

✔ **Interesting Point:** Compounds with the same formula but different molecular structure are called **isomers**. *Iso-* comes from the Greek for "same" or "equal." *Mer* means unit or formula. Isomers are different compounds, even though they have the same formula.

Compound	Boiling Point
n-propanol	97.2°C
isopropanol	82.5°C

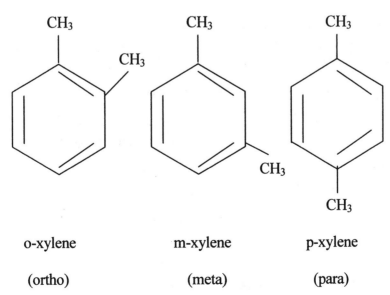

Figure 3.12. Xylene isomers.

XYLENE ISOMERS

Compound	BP	MP
o-xylene	144°C	−25°C
m-xylene	139.3°C	−47.4°C
p-xylene	138°C	14°C

HYDROGEN BONDING

Hydrogen bonds are bonds formed when hydrogen that is covalently bonded to an electronegative atom is attracted to another electronegative atom on another molecule. Hydrogen bonds are *weak* attractions *between molecules*, whereas ionic and covalent bonds are much stronger attractions between atoms. An everyday example of a hydrogen bond is in water (see figure 3.13).

✔ **Key Point:** In water, the oxygen-hydrogen bond is polar, oxygen being the more electronegative element. The molecule is therefore polar. The two lone pairs of electrons on the oxygen atom extenuate this. One end of the molecule is partially negative while the two hydrogen atoms become partially positive. The molecules of water are attracted to one another, with the slightly positive hydrogens attracted to the negative "ends" (the oxygens) of the other water molecules. This intermolecular attraction is termed "hydrogen bonding," and acts almost like a glue holding the molecules of water together. In the case of water, the effect on its physical properties are quite amazing; the boiling point of water, for example, is very much higher than it would be if such bonding did not exist. Otherwise, water would be a gas at room temperature.

When H is bonded to O or N, its lone electron is pulled away from its single proton nucleus. A broken line represents hydrogen bonds because they are *much* weaker than covalent bonds (see figure 3.14).

✔ **Key Point:** Hydrogen bonds are like "glue" between molecules.

Although a single hydrogen bond is very weak, the cumulative effect of an enormous number of hydrogen bonds can significantly affect the properties of the compound.

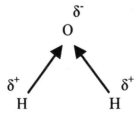

Figure 3.13. Hydrogen bonding.

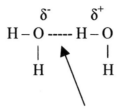

Hydrogen bond

Figure 3.14. Hydrogen bond.

SHORT REVIEW

When an atom combines chemically with another atom, it will do one of the following:

- Gain electrons (become a negatively charged ion or anion)
- Lose electrons (become a positively charged ion or cation)
- Share electrons

✔ **Key Point:** Atomic bonds have characteristic strengths and bond angles that contribute to the shape of molecules.

Chemical Formulas

A chemical formula tells us how many atoms of each element are in the molecule of any substance. A chemical formula uses atomic symbols (from the periodic table) and numerical subscripts to show the relative proportions of the atoms of different elements in a compound.

The chemical formula for water is H_2O. The "H" is the symbol for hydrogen. Hydrogen is a part of the water molecule. The "O" means that oxygen is also part of the water molecule. The "$_2$" after the H means that two atoms of hydrogen are combined with one atom of oxygen in each water molecule. There is no number after the O. When no number appears after a symbol, it means that there is only one atom of that element. The water molecule is made of two atoms of hydrogen combined with one atom of oxygen.

To show the number of molecules, a full-sized number is located *in front* of the molecule. For example 4 molecules of carbon dioxide is designated as $4CO_2$. This means there are a total of 4 C atoms and 8 O atoms in the combination.

Just as in mathematics, we can use parentheses to separate parts in a complex formula. One example is the formula for nitroglycerin, a highly explosive substance. This formula shows that nitroglycerin consists of 3 atoms of C, 5 atoms of H, and the 3 molecules of NO_3. If the parentheses were not used, the formula would not describe the true structure of the nitroglycerin molecule: $C_3H_5(NO_3)_3$.

✔ **Key Point:** Formulas identify the elements, and the number of atoms, that compose the molecule. The formula is a shorthand way of writing *what* elements are present in each molecule of a compound, and *how many* atoms of each element are present in each molecule.

As mentioned, subscripts are important part of chemical formulas. The subscripts in a chemical formula identify the compound. They also tell us the ratios of atoms and molecules in a substance.

✔ **Key Point:** Changing a subscript in the chemical formula changes the identity of the substance. For example, H_2O (water) becomes H_2O_2 (hydrogen peroxide) when we place a subscript 2 on the O atom. On the other hand, placing a number (coefficient) in front of a chemical formula only changes the amount of the substance and not its identity. For example, placing a 2 in front of $2H_2O$ means 2 molecules of H_2O.

WRITING FORMULAS

Determining chemical formulas from diagrams is inconvenient. It is easier to apply the rules shown in table 3.1 to determine, for example, formulas for aluminum oxide.

RADICALS

Some covalently bonded groups of atoms act like single atoms. Though the atoms within a radical are held together by covalent bonds, in each case, they contain either an excess or a deficiency of electrons. Most radicals, however, are negative ions. Table 3.2 gives the names, formulas, and valence numbers of a few of the common radicals.

FORMULA WEIGHTS

The *formula weight* is the sum of the weights of all the atoms in a formula. The formula weight for sodium chloride, NaCl, is found as follows:

Table 3.1. Rules for writing formulas.

Rule	Application	
Write the symbol of the elements.	Al	O
Always write the metal first, the nonmetal second.		
Write the valence of each element above each symbol.	Al^3	O^2
Crisscross the valences.	Al^3	O^2
Reduce the subscripts to their lowest terms.	Al_2	O_3

Table 3.2.   Table of valences.

Metals		
+1	+2	+3
Hydrogen (H)	Magnesium (Mg)	Aluminum (Al)
Sodium (Na)	Strontium (Sr)	Boron (B)
Ammonium (NH_4)	Calcium (Ca)	
Nonmetals		
−1	−2	−3
Fluoride (F)	Oxide (O)	Phosphate (PO_4)
Chloride (Cl)	Sulfide (S)	
Iodide (I)	Carbonate (CO_3)	
Nitrate (NO_3)		

$$
\begin{array}{lr}
\text{Atomic weight of Na} & 23 \\
\text{Atomic weight of Cl} & \underline{35} \\
\text{Formula weight} \quad = & 58
\end{array}
$$

The formula weight of aluminum sulfate, $Al_2(SO_4)_3$ is found as follows:

$$
\begin{array}{lrcr}
\text{Atomic weight of Al } (\times\ 2) & 27 & = & 54 \\
\text{Atomic weight of S } (\times\ 3) & 32 & = & 96 \\
\text{Atomic weight of O } (\times\ 12) & 16 & = & \underline{192} \\
\text{Formula weight} & & = & 342
\end{array}
$$

MOLES

Formula or molecular weight in grams is called a *mole*. The number of moles of a compound is found by using the following expression:

$$
\frac{\text{Actual weight}}{\text{Formula weight}} = \text{Number of moles}
$$

Additional Reading

Smiley, R. A., and H. L. Jackson. *Chemistry and the Chemical Industry: A Practical Guide for Non-Chemists.* Boca Raton, Fla.: CRC Press, 2002.

Summary

- When items share their electrons equally we have *nonpolar covalent bonding*.
- Nonpolar covalent bonding takes place between atoms that have strong and equal attractions for electrons.

- The pairing of one set of electrons forms a single covalent bond.
- The pairing of two sets of electrons forms a double covalent bond.
- The pairing of three sets of electrons forms a triple covalent bond.
- All chemical reactions between atoms involve the exchange or sharing of valence electrons.
- The formula identifies the number of atoms of each element that composes the molecule of a compound.
- A radical is a group of atoms that is covalently bonded together and acts like one atom.

New Word Review

Compound—a material composed of different elements.
Covalent bond—bonding by the sharing of electrons.
Element—a material whose atoms have the same atomic number.
Formula—a combination of symbols that tells how many atoms of each element compose a molecule.
Formula weight—the sum of the atomic weights of the elements present.
Ion—an electrically charged atom.
Ionic bond—the force holding oppositely charged ions together.
Mole—a quantity of a compound equal in weight to the formula weight, or molecular weight, of the compound.
Molecule—the smallest unit of a compound.
Nonpolar—all ends with the same electrical charge.
Polar—oppositely charged ends.
Radical—a group of elements that behaves as a single element. All radicals are ions, possessing an electrical charge.
Subscript—the small number written next to the symbol of an element in a formula. It tells how many atoms of that element are in each molecule.
Valence—the number of electrons an atom gives, takes, or shares when it bonds with other atoms.

Chapter Review Questions

3.1. In the following chemical formulas, determine the number of atoms contained in each formula.

Formula	Element	# Atoms
CO_2	carbon/oxygen	_____
H_2SO_4	hydrogen/sulfur/oxygen	_____
CaO	calcium/oxygen	_____

3.2. The molecules of a compound are composed of atoms of _____ elements.

3.3. The _____ is the smallest unit of a compound.

3.4. Ions with opposite charges will _____ each other.

3.5. Another name for the ionic bond is _____.

3.6. The attractive force of _____ charged ions holds ionic materials together.

3.7. The formula of a compound shows the _____ of each element in its molecule.

3.8. The _____ tells how many atoms of each element are in each molecule of a covalent compound.

3.9. Atoms that are bound together by a _____ covalent have equal attraction for electrons.

3.10. A _____ covalent bond has two nuclei sharing the same electron pair equally.

3.11. A _____ covalent bond is formed between atoms with different attractions for electrons.

3.12. An element that has the symbol F shares _____ electron.

3.13. Compute the formula weight for H_2O.

3.14. Compute the formula weight for HNO_3.

3.15. 100 grams of sodium hydroxide, NaOH, are to be used in making up a solution. How many gram-moles would that be?

3.16. Electrons are found in areas called _____ energy levels.

CHAPTER 4

Inorganic Chemistry and Terminology

Regarding the importance of inorganic chemistry, R. T. Sanderson has written: "All chemistry is the science of atoms, involving an understanding of why they possess certain characteristic qualities and why these qualities dictate the behavior of atoms when they come together. All properties of material substances are the inevitable result of the kind of atoms and the manner in which they are attached and assembled. All chemical change involves a rearrangement of atoms. Inorganic chemistry [is] the only discipline within chemistry that examines specifically the differences among all the different kinds of atoms."[1]

Topics in This Chapter

- Acids, Bases, and Salts
- Metals
- Soils and Minerals

In chapter 1, I pointed out that chemistry is so complex that no one person could expect to master all aspects of such a vast field, so it has been found convenient to divide the subject into specialty areas, for example:

- Organic chemistry
- Inorganic chemistry
- Biochemistry
- Physical chemistry
- Theoretical chemistry
- Analytical chemistry

Traditionally, however, these specialty areas are trimmed to *four* major divisions of chemistry:

Organic: the chemistry of carbon compounds

Inorganic: the chemistry of non-carbon compounds, as well as elemental carbon and simple carbon oxides

Physical: reaction mechanisms, bonding theory, chemical energetic (kinetics), theoretical calculations

Analytical: identification, quantification, separation, and instrumentation (see figure 4.1)

The distinction between organic and inorganic chemistry is often blurred by scientific disciplines such as *organometallic chemistry*. Organometallic compounds contain one or more *carbon-metal* bonds (see figure 4.2), such as a C-Hg bond. Organometallic chemistry is typically thought of as a subset of inorganic chemistry, although it also plays a major role in organic chemistry.

Acids, Bases, and Salts

For the environmental practitioner, acids and bases can be a nightmare. This is especially the case when these substances are mishandled. Acids and bases can cause severe burns to skin, eyes, and respiratory tract. The environmental practitioner must be aware of the engineering controls (ventilation, fume hoods, eyewashes, etc.) installed to make working with acids and bases safe. They must also be aware (i.e., ensure) that all engineering controls are working as per design. Along with knowing and enforcing special handling procedures and storage requirements, the environmental practitioner must also be well versed in decontamination and spill and accident procedures.

In order to understand acids, bases, and salts, we must first be familiar with two important terms, *ionization* and *electrolytes*.

Ionization is the dissociation of ionically bonded compounds into their component *anions* and *cations*.

Electrolytes are compounds which, when dissolved in water, produce a solution of ions that conducts an electric current. There are three types of electrolytes:

Organic	**Physical**
Inorganic	**Analytical**

Figure 4.1. The four major divisions of chemistry.

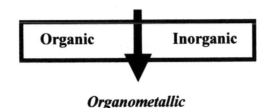

Organometallic

Figure 4.2.   Organometallic compound.

- *Acids*, substances that ionize to produce hydrogen ions (H^+)
- *Bases*, substances that ionize to produce hydroxide ions (OH^-)
- *Salts*, substances that ionize without producing either H^+ or OH^- ions (see figure 4.3)

Strong electrolytes ionize 100% (i.e., completely) in solution; *weak* electrolytes ionize only to a slight extent in solution.

✔ **Key Point:** Always remember that the terms "strong" and "weak" in this context refer to the degree of ionization and never to the concentrations of the electrolytes.

Strong electrolytes are the strong acids: HCl, H_2SO_4, HNO_3 (also the other acid halides) and the strong bases: NaOH, KOH, $Mg(OH)_2$. Water is one of the most important weak electrolytes. It is known that water is 0.00001% (1×10^{-7}).

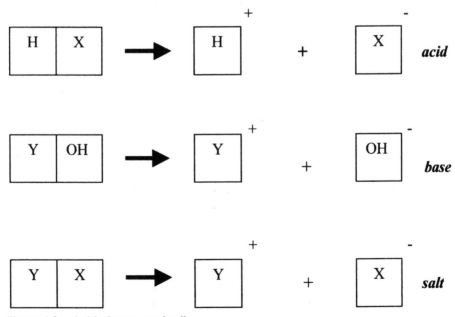

Figure 4.3.   Acids, bases, and salts.

PROPERTIES OF ACIDS

- The strength of an acid depends on how much it ionizes. Hydrochloric acid (HCl) is an acid that ionizes to produce H^+ and Cl^-.
- Acids release a hydrogen ion into water solution.
- Acids neutralize bases in a neutralization reaction. An acid and a base combine to make a *salt* and water. A salt is any ionic compound that could be made with the anion of an acid and the cation of base. The hydrogen ion of the acid and the hydroxide ion of the base unite to form water.
- Acids corrode active metals. When an acid reacts with a metal, it produces a compound with the cation of the metal and the anion of the acid and hydrogen gas.
- Acids turn blue litmus to red. Litmus is one of a large number of organic compounds that change colors when a solution changes acidity at a particular point. Litmus is the oldest known pH indicator. It is red in acid and blue in base.

> ✔ **Interesting Points:** Litmus does not change color exactly at the neutral point between acid and base, but very close to it. Paper impregnated with litmus ("litmus paper,") is commonly found in any drug store, and "litmus test" has moved into popular English for the one factor that determines an outcome.

- Acids taste sour. CAUTION: Tasting acid is strictly prohibited! Acetic acid is the acid ingredient in vinegar. Citrus fruits such as lemons, grapefruit, oranges, and limes have citric acid in the juice.
- *Binary acids* are composed of two elements, hydrogen and a nonmetal. They are named with the prefix *hydro-*, the name of the nonmetal, and the suffix *-ic*.
- Some common acids are HCl (hydrochloric acid), HNO_3 (nitric acid), and H_2SO_4 (sulfuric acid).

PROPERTIES OF BASES

- Bases release a hydroxide ion into water solution.
- Bases neutralize acids in a neutralization reaction.
- Bases neutralize protein. This accounts for the "slippery" feeling on hands when exposed to bases. Strong bases that dissolve well in water, such as sodium or potassium lye, are very dangerous because a great amount of the structural material of human beings is made of protein. Strong bases must be carefully handled to avoid serious damage to flesh.
- Bases turn red litmus to blue.
- Bases taste bitter. Tasting bases is more dangerous than tasting acids due to the property of stronger bases to denature protein. CAUTION: Tasting of lab chemicals is strictly prohibited!
- Sodium hydroxide (NaOH) is a base that ionizes to produce Na^+ and OH^-.
- Almost all bases are composed of a metal and the hydroxide ion. Name the metal first, then the word "hydroxide."

- Some common bases are NaOH (sodium hydroxide), $Ca(OH)_2$ (calcium hydroxide), and $Mg(OH)_2$ (magnesium hydroxide).

PROPERTIES OF SALTS

- Table salt (NaCl) is a salt that ionizes to produce Na^+ and OH^-.
- Salts are composed of the positive half of a base and the negative half of an acid.
- Not all salts are soluble in water.
- A salt is the combination of an anion ($-$ ion) and a cation ($+$ ion).
- In a solid salt, the ions are held together by the difference in charge.
- Salts made of the anion of a strong acid and the cation of a strong base will be neutral salts, that is, the water solution with this salt will have a pH of seven.
- Salt made of the anion of a strong acid and the cation of a weak base will be acid salts; that is, the water solution with this salt will have a pH of less than seven (e.g., ammonium chloride).
- Salts made of the anion of a weak acid and a strong base will be an alkali salt. The pH of the solution will be over seven (e.g., sodium bicarbonate).
- It can be a bit more difficult to tell the pH of a salt solution if the salt is made of the anion of a weak acid and the cation of a weak base.
- Some common salts are KCl (potassium chloride), Na_2SO_4 (sodium sulfate), and H_2SO_4 (sulfuric acid).

OVERVIEW OF pH

The concentration of H^+ in any water solution is normally very important, because H^+ is responsible for acidic properties. Typically, pH is used to express how acidic or alkaline a water solution is. The hydrogen ion concentration is inversely proportional to the hydroxide ion concentration.

By measuring the concentration of hydrogen ions (H^+) in a solution we can determine whether a solution is an acid or base. While the concentration can be expressed in powers of 10, it is more conveniently expressed as pH. Pure water, for example, has 1×10^{-7} grams of hydrogen ions per liter. The negative exponent of the hydrogen ion concentration is called the pH of the solution. The pH of water is 7—a neutral solution. A concentration of 1×10^{-12} has a pH of 12. A pH less than 7 indicates an acid solution, and a pH greater than 7 indicates a basic solution.

✔ **Interesting Point:** In pH, the "p" is for the German word *potentz* and the H stands for hydrogen.

Remembering that pure water itself ionizes to a very small extent, scientists developed the *pH scale* to work with these very small concentrations as whole numbers. Based on negative logarithms, the pH of pure water at room temperature is 7 (i.e., neutral

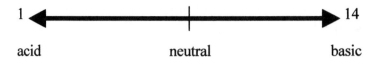

Figure 4.4. The pH scale.

pH). A pH less than 7 is indicative of an acidic solution. A pH greater than 7 is indicative of a basic solution (see figure 4.4).

✔ **Key Point:** Because the pH scale is a logarithmic scale, the difference between each number on the scale (e.g., 6 and 7) indicates a tenfold change in the hydrogen ion concentration. A solution of table sugar in water is neutral (pH 7) because it does not contain hydrogen ions nor does it react with bases to produce water.

Control of pH is of critical importance in many industrial operations such as water purification and wastewater treatment.

Metals

Metals can also cause problems in the workplace. The person responsible for EH&S, for example, must have an emergency cleanup and spill containment plan for mercury spills and their mitigation.

Of the 100+ known elements, more than 86 are metals. According to the periodic table, metals are listed as *alkali metals* (Group I), the *alkaline earth metals* (Group II), *iron and steel alloy metals*, *nonferrous metals*, and the *noble and rare metals*. Metals are defined as a substance that is malleable and ductile, has a characteristic luster, and is generally a good conductor of heat and electricity. In addition to the characteristics just mentioned, metals exhibit other common physical and chemical properties. These common properties are listed in table 4.1.

COMMON PROPERTIES OF METALS

Most metals are very dense solids because the atoms are closely packed together in a regular arrangement called a *lattice* or *crystal* structure (see figure 4.5). Metals are often

Table 4.1. Common properties of metals.

Physical Properties	Chemical Properties
Solid (except Hg)	Form lattices
Dense (heavy)	Form cations
Malleable	Form oxides
Ductile	Form salts
Good conductors	Form hydroxides
Shiny (luster)	

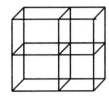

Figure 4.5.   Lattice and crystal structure typical of dense metals.

soft and easily bent (malleable) or distorted because the lattice structure allows the layers of atoms to slide over each other (see figure 4.6). Sometimes the soft nature of metals is useful, such as gold being beaten into a film and copper being drawn into wires (an example of *ductility*).

To make metals stronger and less easily distorted, chemists add specific amounts of other elements (usually metals, such as titanium, chromium, cobalt, nickel, silicon, zirconium, and others) to form *alloys*. For example, steel is an alloy of iron and carbon, with other metals added to give it special properties. Other common alloys include brass, which is composed of 60% copper (Cu) and 40% zinc (Zn), and bronze, which is composed of 90% copper (Cu) and 10% tin (Sn). These alloy metals are added to metals to increase their strength, corrosion resistance, magnetism, hardness, flexibility, and other special properties.

✔ **Key point:** Specific amounts of other metals added to form alloys effectively form a "wedge" that makes the metal stronger.

Metals generally conduct heat and electricity well because the *outer shell electrons* of the atoms in the lattice become detached and form a "sea" of electrons on the surface

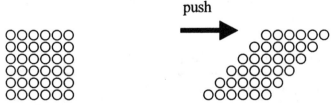

Figure 4.6.   Metal lattice structures are easily bent or distorted by layers of atoms that slide over each other.

of the metal. Copper and aluminum are two common metals used to conduct electricity. Metal radiators heat houses by conducting heat.

The highly mobile "sea of electrons" on the surface of metals also explains their high luster. When a light beam strikes the surface of a metal, it causes the mobile electrons to oscillate back and forth. Like any moving electric charge, these oscillating electrons give off energy in the form of light; the net effect is that the light beam is reflected.

CHEMICAL REACTIVITY OF METALS

The *chemical reactivity* of metals increases with increasing tendency to *ionize* (see figure 4.7). Most metals react with oxygen to form *oxides* as shown here:

$$K + O_2 \rightarrow KO_2 \text{ (vigorous)}$$
$$Cu + O_2 \rightarrow 2CuO \text{ (slow)}$$
$$Ag + O_2 \rightarrow \text{ (no reaction)}$$
$$Au + O_2 \rightarrow \text{ (no reaction)}$$

Metals react with *acids* to form *salts* and release hydrogen, as shown here:

$$2Na + 2HCl \rightarrow 2KOH + H_2 \uparrow \text{ (vigorous, fire)}$$
$$Mg + 2HCl \rightarrow MgCl_2 + H_2 \uparrow \text{ (moderate)}$$
$$Pb + 2HCl \rightarrow PbCl_2 + H_2 \uparrow \text{ (very slow)}$$

Metals react with *water* to form *hydroxide*, as shown here:

$$2K + 2H_2O \rightarrow 2KOH + H_2 \uparrow \text{ (vigorous, fire)}$$
$$2Na + 2H_2O \rightarrow 2NaOH + H_2 \uparrow \text{ (vigorous, no fire)}$$
$$Ca + 2H_2O \rightarrow Ca(OH)_2 \downarrow + H_2 \uparrow \text{ (slow)}$$

METAL NOMENCLATURE

In chemical formulas for inorganic compounds, the most *electropositive* constituent (cation) is generally named first without modification (see figure 4.8).

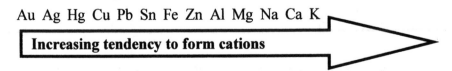

Figure 4.7. Chemical reactivity of metals.

Rn Xe Kr B Si C Sb As P N H Te Se S At I Br Cl O F

⟶

Named first   Named last

Figure 4.8. Metal nomenclature.

✔ **Key Point:** IUPAC (International Union of Pure and Applied Chemistry) has established a priority sequence for binary compounds between metals; the highest priority element is named first (XeF_2, NH_3, and H_2S).

In general, prefixes like *mono, di, tri,* etc. usually indicate the number of attached groups. Examples are:

$W(CO)_6$	tungsten *hexa*carbonyl
$SiCl_2H_2$	*di*chlorosilane
$[Co(NH_3)_4Cl_2]^2$	tetraaminedichlo rocobalt (III)

In some cases, other prefixes such as *bis-, tris-, tetrakis-,* etc. are used to avoid ambiguity. An example is trisdecyl phosphine (to avoid confusion with the tridecyl "radical" $C_{13}H_{27}$).

$$P(C_{10}H_{21})_3$$

If the electronegative constituent is monatomic, its name is modified to end in *-ide* (e.g., carbon disulfide, CS_2). If the electronegative constituent is heteropolyatomic, its central atom is modified to end in *–ate* (with some exceptions); an example is:

$K_3[Fe(CN)_6]$ potassium hexacyanoferr*ate* (III)

In IUPAC nomenclature, the names of *monoatomic cations* are not modified.

Cu^+	the copper (I) ion
Cu^{+2}	the copper (II) ion

Historically, the cation of higher oxidation state was given an *-ic* ending, while the cation of lower oxidation state was given the ending *-ous*.

Br_2Cu	cupr*ic*
BrCu	cupr*ous* bromide
Cl_3Fe	ferric chloride
Cl_2Fe	ferrous chloride

The names of *monoatomic anions*, as well as some polyatomic anions, are modified to end in *-ide*.

H$^-$	hydride ion
OH$^-$	hydroxide ion
F$^-$	fluoride ion
CN$^-$	cyanide ion

Historically it has been customary to indicate the presence of oxygen by a series of prefixes and sometimes the suffix –*ite*.

NO^{2-}	nitrite
N$_2$O$_2$$^{-2}$	hyponitrite
NOO$_2$$^{-2}$	peroxonitrite

Metals often form *coordination compounds* (or complex ions) that contain one or more *coordinate covalent bonds*, that is, compounds in which ions and/or neutral molecules are attached to metal ions. In a *coordinate covalent bond*, both of the electrons of the bond come from the same atom. The *coordination number* of the central atom (typically the metal cation) is the number of atoms that are directly linked to it.

MnO$_4$$^-$	Coordination number = 4
Fe(CO)$_5$	Coordination number = 5

The atoms, radicals, or molecules attached to the central metal atom are called *ligands*. Ligands may be either ions or neutral molecules. Within a ligand, the atom that attaches directly to the metal by a coordinate covalent bond is called the *donor atom*. Common ligands are:

• The halides: F$^-$, Cl$^-$, Br$^-$, I$^-$
• The sulfide ion: S^{-2}
• The hydroxide ion: OH$^-$
• The cyanide ion: CN$^-$
• The nitrite ion: NO$_2$$^-$

✔ **Interesting Point:** The colors of copper-bearing minerals are due to differing ligand field splitting of the copper ions' orbitals.

✔ **Interesting Point:** Alfred Stock first used the term ligand (from *ligare* [Latin], to bind) in 1916 in relation to silicon chemistry. It didn't appear in a British journal until H. Irving and R. J. P. Williams used it in *Nature*, in 1948 (pp. 162, 746).

Parentheses ( ) and brackets [ ] are used to enclose identical sets of atoms. Brackets are typically used to close off *complex* ions or neutral *coordination entities*.

Arabic numbers are used as subscripts to indicate the number of atoms and as prefixes to indicate the position of the substitution or addition in a chain. *Roman numerals* are used in ( ) to indicate the *oxidation number* of an element.

The oxidation number is the charge that would be present if the electrons of each bond were assigned to the more electronegative atom.

NaCl1 Na^+ and 1 Cl^-
                               $Na = 1$ and $Cl = -1$
$Mg(OH)_2^-$ 1 Mg^{+2}, 2O^{-2}, and 2H^+
                               $Mg = II$, 2$O = -II$, 2$H = 1$

In chemical formulas for coordination compounds, the central atoms(s) is/are usually placed first (except in structural formulas). In the *names* of coordination compounds, the central atom(s) is/are placed *after* the names of the ligands.

$K_3[Fe(CN)_6]$ potassium hexacyanoferrate (III)

ALKALI METALS

The elements of Group I, the alkali metals, are:

- Lithium (Li)
- Sodium (Na)
- Potassium (K)
- Rubidium (Rb)
- Cesium (Cs)
- Francium (Fr)

This group takes its name from the fact that they all react with cold water to form *alkaline* (i.e., basic) solutions. All these metals are soft and silvery. They can be easily cut with a knife to expose a shiny surface, which dulls on oxidation. They are good electrical and thermal conductors. They emit electrons when struck by light. When heated, they emit colored light.

✔ **Interesting Point:** When heated, lithium turns red, sodium turns yellow, potassium turns red, and cesium turns blue.

These metals are highly reactive elements because they easily lose their single outer shell electron to become cations. They are not found uncombined in nature. They have only one oxidation state: $+1$. The reactivity increases on descending the group from lithium to cesium. There is a closer similarity between the elements of this group than in any other group of the periodic table.

Alkali metals react explosively with water to release H(g). They react violently with dilute acids to release H(g). When exposed to air they tarnish and burn vigorously when heated.

Sodium (Na) and potassium (K) are the sixth and seventh most abundant elements on Earth, respectively. The other members of this group are relatively rare.

✔ **Interesting Point:** Seawater = 1.14% Na and 0.04% K.

Sodium is found in vast salt deposits (NaCl) and other sodium salts ($NaNO_3$, Na_2SO_4, Na_2CO_3, $Na_2B_4O_7$), especially in arid regions. Sodium is also found in silicate rocks. Potassium is found in large KCl and K_2CO_3 beds, as well as silicate rocks. Other important alkaline metal compounds include:

- NaOH (sodium hydroxide, caustic soda)
- $NaHCO_3$/Na_2CO_3 (sodium carbonate/bicarbonate)
- Various Na salts of organic compounds
- KNO_3 (potassium nitrate, saltpeter)
- KOH, K_2CO_3 (potash)

THE ALKALINE EARTH METALS

The elements of Group II, the alkaline earth metals, are:

- Beryllium (Be)
- Magnesium (Mg)
- Calcium (Ca)
- Strontium (Sr)
- Barium (Ba)
- Radium (Ra)

These metals were named centuries ago for the "earthy" nature of the compounds then known (e.g., Fe_2O_3 and Al_2O_3) and their tendency to form alkaline suspensions in water.

✔ **Interesting Point:** These compounds were mistaken for elements until the 1800s, by which time the name for the group was already well established.

The alkaline earth metals are similar to the Group I metals; they are soft, white metals (except Be). However, they are harder and denser than sodium and potassium and have higher melting points. They are good conductors of heat and electricity. Barium is a good electron emitter. Ca, Sr, and Ba emit colored light upon heating (Ca = brick red, Sr = brilliant crimson, and Ba = green). They are all found in Earth's crust, but not in the elemental form, as they are very reactive. Instead, they are widely distributed in rock structures. The minerals in which magnesium is found are magnesite and dolomite. Calcium is found in chalk, limestone, gypsum, and anhydrites. Magnesium is the eighth most abundant element in Earth's crust, and calcium is the fifth (approximately 3.6% of the crust). The remaining alkaline earth metals are found in trace quantities.

✔ Marie and Pierre Curie discovered the radioactive element Ra in 1898.

The alkaline earth metals, as with Group I metals, are very reactive because they readily lose their 2 outer shell electrons to form +2 cations. They are not found in the free state in nature. They all have an oxidation state of +2.

All of these metals react with water to release hydrogen (except Be), tarnish immediately in air, and burn vigorously when heated.

These metals are most often found as carbonates, silicates, and sulfates. Calcium is naturally found in many forms, including limestone, marble, chalk, seashells, pearls, and gypsum. Magnesium also occurs in various forms, such as Epsom salts and talc.

✔ **Interesting Point:** Both Ca and Mg are found in asbestos and dolomite.

Other important alkaline earth compounds include:

- Cement
- Lime
- Magnesium hydroxide
- Barium sulphate
- Emerald and aquamarine

ALUMINUM, TIN, AND LEAD

Aluminum (Al), which has many of the chemical and physical properties of the alkaline earth metals, is easily the most abundant metal on earth, making up more than 7% of Earth's crust. It is found in various silicate rocks, including mica $KH_2Al_3(SiO4)_3$) and feldspar ($KAISi_3O_8$).

Because of aluminum's light weight, high tensile strength, and resistance to corrosion, it has been favored as a building material and for containers. It is also used in the manufacture of paints, dyed fabrics, bricks, and ceramics. Rubies, sapphires, amethysts, and topazes are made of Al_2O_3 crystals with various metal impurities.

The chief source of tin (Sn) is cassiterite (SnO_2), found in readily accessed deposits in Malaysia and Bolivia. Tin is a soft, white metal with a low melting point and typical valences of +2 and +4.

In recent times, there were more organometallic tin compounds in industrial use than any other metal, with an annual worldwide production in excess of 40,000 metric tons. The principal use of tin is in the electrolytic production of tin plate, sheet iron coated with tin to protect against corrosion.

✔ **Interesting Points:** Specimens of tin found in Egyptian tombs bear witness to its long use by man. Coating thin sheets of mild steel with tin produces "tin cans." "Tin disease" arises from the existence of two allotropic forms of white and gray tin.

Tin biocides were once used to prevent the growth of organisms on the undersides of boats, preserve wood, leather, paper, and textiles, and stop fungal growth in cooling water towers.

Lead (Pb) is a soft metal that has little tensile strength and the greatest density of the common metals (except for mercury and gold). Lead is primarily obtained from galena (PbS). Because of its high density and very slow rate of interaction with many substances, it is often used for lining containers or shielding. Lead forms a key component of batteries used in cars. Lead is an important component of solder. Though now discontinued, lead was once used in lead-based paints, lead gasoline additives, and lead ceramic glazes.

The ancient Romans used lead for making water pipes, cooking utensils, water tanks, and storage vessels. Many scientists and historians have concluded that the resulting lead contamination was a major cause of the decline of the Roman Empire. Mosonius, a Roman writing in the first century A.D., observed that masters were physically weaker than the servant class, country people were stronger than city folk, and people who ate plain food (and less lead-laced wine) were likely to live longer and have fewer diseases associated, by hindsight, with lead poisoning—"gouts, dropsies, and colics." This is as close as anyone got to discovering chronic lead poisoning in the Roman Empire and leaving a record of the hypothesis.

It is evident that rich Romans received more than their share of lead poisoning because they could afford more of the sources of lead contamination. For example, the rich controlled most of the public water outlets and demanded the first drawn water of the morning—which had been sitting over night absorbing lead. The evidence suggests that the offspring of parents with lead poisoning were more likely to be underachievers or not survive infancy. Thus chronic lead poisoning persistently destroyed the Roman aristocracy and its ability to rule the empire.

TRANSITION METALS

The transition metals are generally defined as the "d block elements," i.e., those elements in which d orbitals are being filled. Some more exact definitions exclude scandium and zinc, which do not follow the same trends as the other transition metals.

In general, these elements, which act as catalysts, are all hard metals with high melting and boiling points. They exhibit multiple oxidation states. Transition metals typically form colored ions and compounds and tend to form complex ions (coordination compounds). The important transition metals include:

- Fe
- Cu
- Ag, Au, Pt
- Cr, Hg, Cd
- Trace Elements

Iron is second only to aluminum in the list of abundant metals on Earth. Cast iron is heavy, brittle, not malleable, and not ductile, cannot be welded, rusts easily, and must be cast into molds. It was one of the first metals that humans learned to extract from ore and forge into tools.

Steel is formed from specific combinations of iron, carbon, and/or other transition metals. Compared to cast iron, steel is lighter, tougher, malleable, ductile, can be welded, can be shaped, and is more resistant to rust.

Copper (Cu) is primarily obtained from its ores, although some free copper occurs naturally. It is used in copper wires and tubing, as well as in copper alloys such as brass (Cu/Zn), bronze (Cu/Sn/Zn), and dental gold (Au/Cu).

Silver (Ag), *gold* (Au), and *platinum* (Pt) are known as the noble metals (sometimes called native metals) because they occur naturally in an uncombined state and do not readily react with common reagents. Silver is the best conductor of heat and electricity. Gold is the most malleable and ductile of the metals. Platinum does not tarnish and is easily worked.

Silver bromide (AgBr) and other silver salts are used in photography. Visible light has sufficient energy to remove an occasional electron from $Br-$. This electron migrates to $Ag+$, forming a silver atom that acts as a nucleus for forming grains of silver visible to the naked eye.

MERCURY (Hg)

Mercury (quicksilver) is the only metal that is a liquid at room temperature.

It is most commonly found as the sulfide *cinnabar* (HgS). Because of its chemical inactivity, mobility, high density, and electrical conductivity, it is used extensively in barometers, vacuum pumps, and liquid seals, and for electrical contacts. Mercury forms alloys called *amalgams*, which are used in dentistry.

TRACE ELEMENTS

Some of the transition elements are referred to as trace elements, because they are essential to the human diet in minute amounts. Trace elements include:

- Cobalt in vitamin B_{12} (development of red blood cells)
- Copper in cytochrome oxidase (aerobic respiration) and tyrosinase (melanin production)
- Manganese in phosphates (bone development)
- Molybdenum in plant nitrate reductases (amino acid synthesis)
- Zinc in alcohol dehydrogenase (alcohol degradation), carbonic anhydrase (CO_2 transport), and carboxypeptidase (protein digestion)

THE LANTHANIDES AND ACTINIDES

The lanthanide series (rare earth elements), beginning with cerium (element 58) and extending through lutetium (element 71), are very similar in chemical and physical properties to each other. The series of elements beginning with thorium (element 90) and extending through lawrencium (element 103) is called the *actinide series*. All the elements of this series are radioactive. These rare metals are also referred to as the *inner transition metals* because the "inner" *f* orbitals are being filled in both series. Uranium is the most abundant of these elements. Elements with atomic numbers higher than 92 are all manmade.

✔ **Interesting Points:** The lanthanides and actinides are very difficult to separate because they differ in the third outermost shell. U, Th, and Pu are used as sources of atomic energy. Eu_2O_3 (a brilliant red) is used in TV receivers. Sunglasses contain La_2O_3 to protect the eyes from UV radiation. The uranium isotopes needed for the first atomic bomb were separated using $UF_6(g)$.

THE METALLOIDS: SILICON AND GERMANIUM

Some elements, called metalloids, or semi-metals, cannot be satisfactorily identified as being either metals or nonmetals, for they possess some of the properties of each. Silicon (Si) and germanium (Ge) are metalloids.

Very pure samples of silicon and germanium are nonconductors because the valence electrons are "locked up" in four bonds to each atom's neighbor (see figure 4.9).

If as little as 0.001% of As or B is introduced (see figure 4.10), the conductivity of the "doped" crystal structure is dramatically increased. An atom of arsenic (As) can fit into the crystalline lattice of germanium if it gives up its fifth valence electron. This free electron can move through the crystal structure under the influence of an electric field (an n-type semiconductor; see figure 4.10).

An atom of B can also fit into crystalline germanium, but since it has only 3 valence electrons, it creates an electron *deficiency* in the lattice (a p-type semiconductor).

Semiconductors with a junction between electron-rich and electron-poor regions act as a *rectifier*, which changes alternating current into direct current. These properties make these two elements ideally suited for making computer chips and electronic devices.

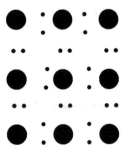

Figure 4.9. The metalloids: Si and Ge.

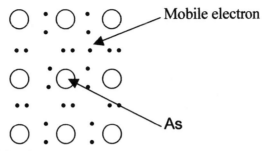

Figure 4.10. Mobile electron in an n-type semiconductor.

Soils and Minerals[2]

The EH&S professional responsible for environmental compliance must have some knowledge of soils and soil chemistry. Soil should be kept "clean" because the complex web of animal and plant life on earth depends on a thin layer of clean soil.

The idea of "clean" soil is understandably foreign to most people, because soil pollution is seldom visible and the extent of soil pollution and its effects is not common knowledge. Soil is complex—so complex that it's hard to know what should or shouldn't be in it. To understand the idea of soil pollution, you need to look at soil from the standpoint of its function. Soil can be defined as the topmost layer of decomposed rock and organic matter which usually contains air, moisture, and nutrients and can therefore support life. Most people would have little difficulty in understanding and accepting this simple definition. Why, then, is there so much confusion about the exact meaning of the word *soil*? Quite simply, it is because soil is not simple but quite complex. Like the term *pollution*, the exact definition of soil is a personal judgment call. For example, some people think of soil in terms of dirt.

First of all, soil is not dirt. Dirt is misplaced soil—soil where we don't want it, such as on our hands, clothes, automobiles, or floors. Dirt we try to clean up and to keep out of our living environments. Secondly, soil is too special to be called dirt because soil is mysterious and essential to our existence—there is nothing mysterious about dirt and we certainly do not need it.

From an EH&S standpoint, we cannot afford to treat soil like dirt—degrade it, abuse it, throw it away, contaminate it, or ignore it.

So, what is soil? Soil is composed of clay, air, water, sand, loam, and organic detritus of former life-forms. If water is Earth's blood and air is Earth's breath, then soil is its flesh and bone and marrow.

Soil has five main functions important to life on Earth: (1) soil is a medium for plant growth; (2) soils regulate our water supplies; (3) soils are recyclers of raw materials; (4) soils provide a habitat for organisms; and (5) soils are used as an engineering medium.

Any fundamental discussion about soil should begin with a definition of what soil is. The word *soil* is derived through Old French from the Latin *solum*, which means floor or ground. A more generalized definition of soil is provided by *The Soil*

Science Society of America in Washington, D.C.: "Soil is unconsolidated mineral matter on the surface of the earth that has been subjected to and influenced by genetic and environmental factors of parent material, climate, macro- and microorganisms, and topography, all acting over a period of time and producing a product—soil—that differs from the material from which it is derived in many physical, chemical, and biological properties and characteristics."

SOIL PROPERTIES

Soil produces most of the food for living things on Earth. Typically, soil is composed of 50% inorganic matter and 5% organic matter. The inorganic component of soil is derived from the weathering of rocks. Therefore soils are composed of the most common elements in Earth's crust: Oxygen, silicon, aluminum, iron, calcium, sodium, potassium, and magnesium. High proportions of organisms such as bacteria, fungi, and earthworms are also found in soil. Approximately 45% of typical soil volumes arise from air-filled pores. The oxygen content of this soil may be as low as 15% due to biodegradation of organic matter. The physical characteristics of soil are largely determined by soil particle size, which ranges from 0.002 mm (clay) to > 1 mm (gravel). Soils are typically found in distinct layers called *horizons*.

A Horizon/Topsoil: layer of maximum biological activity containing most of the organic content; considerable leaching of metal ions and clay particles occurs.

B Horizon/Subsoil: receives leached materials (organics, salts, clay particles) from topsoil.

C Horizon: weathered parent rocks.

D Horizon: bedrock.

Common soil minerals include finely divided quartz, oxides of manganese and titanium, calcium and magnesium carbonates, geotite, magnetite, albite, and orthoclase.

Soils also provide natural plant nutrients such as nitrogen, phosphorus, and potassium. Soil minerals also provide essential micronutrients (boron, chlorine, copper, iron, manganese, molybdenum, sodium, vanadium, and zinc).

The most significant organic component of soils is derived from the partial decay of plant matter. This organic matter not only has high cation affinity but also affinity for low-solubility substances such as DDT. Stormwater runoff, loaded with organic soil components, may cause problems in wastewater treatment.

✔ **Important Point:** Approximately one-third of all U.S. topsoil has been lost to erosion since cultivation first began on this continent. Perhaps as much as 48 million acres of cultivated land is thought to be eroding at unacceptable rates (i.e., > 14 tons of topsoil/acre/year).

MINERALS

A mineral is a naturally occurring, inorganic, homogeneous solid with a characteristic chemical composition, and ordered atomic arrangement. In order for a substance to be labeled a mineral all of these criteria have to be true. Specifically:

Naturally occurring: Formed by natural processes (not created in a lab).

Homogeneous solid: A mineral consists of a single solid compound or element, which cannot be physically subdivided into other chemical components.

Inorganic: Not created by organisms or from their remains. This qualification excludes those substances that are solely a product of organic activity (e.g., sugar), and not substances that are commonly inorganic but may be precipitated by organisms (i.e., calcite).

Characteristic chemical composition: This signifies that a mineral can be represented by a specific chemical formula.

Ordered atomic arrangement: The atoms forming a mineral are arranged in a geometric pattern, which persists throughout the entire mineral. All minerals by definition are crystalline. (Note: Substances in which atoms are randomly arranged are called *amorphous*. All glasses by definition are amorphous.)

✔ **Key Point:** A mineraloid (e.g., bauxite—major ore of aluminum) is a solid which has all the properties of a mineral except that it has no ordered atomic arrangement and has a slightly variable chemical composition.

Additional Reading

Inorganic Chemistry Nomenclature: Perma-Chart (10-pack). Papertech Marketing Group Incorporated, 2000.

SUMMARY

- Acids are compounds that produce hydronium ions (H_3O^+) in water.
- Strong acids and bases are completely ionized in water.
- Weak acids and bases are partly ionized in water.
- A base or alkaline substance produces hydroxide ions $(OH)^-$ in water.
- Bases are bitter. Acids taste sour.
- Bases and acids are detected with indicators. Acids turn litmus red and bases turn litmus blue.
- Acidity is rated in pH units. Acids have a pH range less than 7. Bases have a pH range that is more than 7. Neutral solutions have a pH that is equal to 7.
- When the pH changes by one whole number, the acidity changes ten times. A pH of 6 is ten times more acidic than a pH of 7.
- Acids and bases neutralize each other to form a salt and water.
- About 80% of the elements in the universe are metals.
- Metals have different activities.
- Cast iron is an impure form of iron obtained from the blast furnace.
- Steel has many advantages over cast iron.
- Steel is an alloy of iron and carbon, which controlled impurities added for special properties.

- The chief ore of aluminum is bauxite.
- The alkali metals are soft, light metals. Having only one electron in their outermost shell, they are extremely active.
- The oxides of alkaline earth metals form mildly alkaline solutions.
- Copper is second to iron in importance among the metals.
- The principal alloys of copper are brasses and bronzes.
- The noble metals occur free in nature.
- Typical productive soil is composed of 95% inorganic matter and 5% organic matter.
- Rocks are mixtures of minerals.

✔ **Important point:** Inorganic chemistry is the chemistry of noncarbon compounds, as well as elemental carbon and simple carbon oxides.

New Word Review

Acid—a compound that produces hydrogen ions in water. The hydrogen ions react with the water to make hydronium $(H_3O)^+$.

Base—a compound that produces hydroxide ions in water.

Bronze—an alloy of copper and tin.

Catalyst—a substance that changes the speed of a chemical reaction without affecting the yield or undergoing permanent chemical change.

Coordinate covalent bond—a bond formed when one atom provides both electrons in a shared pair.

Coordination compound—a molecule or ion formed by the bonding of a metal atom or ion to two or more ligands by coordinate covalent bonds.

Ligand—an ion or neutral molecule attached to the central metal ion in a coordination compound.

Metal—a substance that is malleable and ductile, has a characteristic luster, and is generally a good conductor of heat and electricity.

Mineral—a natural chemical compound.

Ore—a mineral-bearing rock from which a metal is extracted.

pH—a scale of 0–14 used for measuring the acidity of a solution.

Salt—a combination of a metal and a nonmetal.

Chapter Review Questions

4.1. In any neutralization reaction, an acid plus a base will produce a salt plus:
    a. water
    b. oxygen gas
    c. hydrogen gas
    d. another salt

4.2.   The formula for sulfuric acid is:
    a. HCl
    b. HNO_3
    c. H_2SO_4
    d. H_2CO_3

4.3.   The acid found in vinegar is:
    a. sulfuric
    b. acetic
    c. hydrochloric
    d. nitric

4.4.   To test for acid, use:
    a. limewater
    b. blue litmus
    c. a potted plant
    d. red litmus

4.5.   Which of the following oxides when dissolved in water will form a base?
    a. CO_2
    b. SO_2
    c. CaO
    d. P_4O_8

4.6.   As the strength of an acid increases, the strength of its conjugate base:
    a. increases
    b. decreases
    c. remains the same

4.7.   The pH of a solution with a H+ concentration of 3.2×10^{-4} is:
    a. 3.5
    b. 7.0
    c. 9.0
    d. 11.3

4.8.   Which of the following is *not* a property of an acidic solution?
    a. changes blue litmus to red
    b. slippery feel
    c. has a pH below 7
    d. sour taste

4.9.   A form of iron that cannot be welded is _____.

4.10. An _____ is a natural mixture from which a valuable metal is extracted.

4.11. Heating a solution of ammonium hydroxide liberates the gas _____.

4.12. Why is lead unfit for use as pipes in drinking water?

4.13. The most common ion in Earth's crust is _____.

4.14. You are given an unknown mineral to identify. You perform a hardness test by scratching it with your fingernail. You are able to leave a scratch mark on it, so you conclude that _____.

Notes

1. R. T. Sanderson, *Simple Inorganic Substances: A New Approach* (Krieger, 1989).

2. This section is adapted from Frank Spellman's *The Science of Environmental Pollution* (Lancaster, Pa.: Technomic Publishing Company, 1999), 35–70.

CHAPTER 5

Organic Chemistry and Terminology

The branch of chemistry that deals with carbon compounds is called *organic chemistry.* The name "organic" was derived from the fact that carbon compounds were originally obtained from living things or the remains of living things (e.g., coal). Organic versus inorganic is a matter of definition rather than source.

Topics in This Chapter

- Organic Nomenclature
- Organic Chemistry
- The 411 on DDT

When we consider that only one out of every thousand atoms on the surface of Earth is carbon, we have to wonder why an entire branch of chemistry is devoted to such a rare element. Carbon is a rare element, but it forms more different compounds than all other elements put together. We can also say that carbon is a special element because it forms the basis of all living things. Carbon is found in all foods. Its unique ability to bond itself in almost endless chains, designs, and shapes also makes it special.

Before 1828, the term *organic* was used to describe compounds occurring in or derived from living organisms. Such substances as starch and urea were classified as organic, for living plants produce starch, and urea is contained in urine. However, in 1828, Friedrich Wohler made urea from ammonium cyanate. Thus, it was proven that urea could be produced, along with several thousand other carbon compounds, from chemicals.

The study of carbon chemistry, or organic compounds, in the modern sense, is the study of compounds of carbon that contain either carbon-carbon bonds or carbon-hydrogen bonds or both. Simply, organic chemistry is a unique field of chemistry because of the specific properties of the carbon atom.

In terms of environmental investigations, the key to organic chemistry is recognizing key functional groups. The whole language of organic chemistry is based on nomenclature and basic structures. The more we learn these basics, the easier it will be to understand how chemical reactions affect them. If we can easily pick out ketones, alcohols, carboxylic acids, aldehydes, ethers, amines, and hydrocarbons (alkanes, alkenes, alkynes, and cyclic hydrocarbons) and name them, we will be in a very good position to start understanding how and why reactions occur.

Accordingly, in this chapter I present organic chemistry nomenclature first—to facilitate understanding—before covering the basics of organic chemistry in general.

Organic Nomenclature

The nomenclature of organic chemistry originates from various sources. The names of organic compounds came about in a variety of ways, including the historical or trivial. Compounds often acquired *historical names* before their exact structure was known. These compounds were sometimes named for their source of origin, the chemist who discovered them, or their friends and relatives.

Unfortunately, there can be different historical (trivial) names for the same compound. Consider, for example, methanol (CH_3OH). Methanol is a clear, colorless liquid with a pungent odor at ambient temperatures. It can cause permanent blindness and death from massive exposures. Synonyms for methanol include:

• Methyl alcohol
• Carbinol
• Wood alcohol
• Wood spirits
• Columbian spirits
• Colonial spirit
• Methyl hydroxide
• Wood naphtha
• Pyroxylic spirit

Some historical names give chemists hints about the compound's structure, but others do not (emphasized segments below give a hint):

• *Meth*anol
• *Methyl alcohol*
• Carbin*ol*
• Wood alco*hol*
• Wood spirits
• Colonial spirits
• Columbian spirits

To make matters more confusing, chemical manufacturers also use *trade names* to create brand recognition. Consider the following examples.

• Clorox (ClNaO)
• Teflon (PTFE)
• Nicoderm (nicotine)
• Aroclor (a PCB: polychlorinated biphenyl)

Modern practices for naming organic chemicals began in the late nineteenth century when IUPAC chemists met in Geneva to develop a formal system of chemical nomenclature. IUPAC's *Nomenclature of Organic Chemistry* set up a system so designed that the *structure* of each organic compound can be used to derive a unique chemical name.

In the following, I provide an overview of IUPAC nomenclature for the compounds that are to be presented later:

Hydrocarbons	Acids
Aromatics	Esters
Halogens	Ethers/Peroxides
Alcohols	Nitriles
Aldehydes	Amines
Ketones	Nitro compounds

HYDROCARBONS (ALKANES)

Alkanes are hydrocarbons that contain no double or triple bonds (see figure 5.1). IUPAC names for alkanes end in *–ane*. Alkane nomenclature is shown in figure 5.2.

In summary, the IUPAC rules for naming alkanes are to:

1. Identify the *longest* alkyl chain and number so that any substituents (branches) receive the *lowest* number(s) possible.
2. Identify each branch/substituent and its number.
3. Attach the name(s) and number(s) of the branch(es) or substituent(s) to the name of the parent (see figure 5.4).

✔ **Key Point:** If more than one substituent is present, the substituent with the highest priority receives the *lowest* number (i.e., the "best" is first; see figure 5.5). IUPAC has arbitrarily established a system of priorities for substituents.

HYDROCARBONS (ALKENES AND ALKYNES)

Hydrocarbons with one or more *double* bonds are called alkenes, and the IUPAC ending is changed to *-ene* (see figure 5.6). Hydrocarbons with one or more *triple*

```
    H H H H H
H---|---|---|---|---|---H
    H H H H H
```

$$CH_3CH_2CH_2CH_2CHCH_2CH_2CH_3$$
$$|$$
$$CH_2CH_3$$

Pentane   **4-ethyloctane**

Figure 5.1. Alkanes.

CH$_4$	methane
CH$_3$CH$_3$	ethane
CH$_3$CH$_2$CH$_3$	propane
CH$_3$CH$_2$CH$_2$CH$_3$	butane
CH$_2$CH$_2$CH$_2$CH$_2$CH$_3$	pentane
CH$_3$CH$_2$CH$_2$CH$_2$CH$_2$CH$_3$	hexane
CH$_3$CH$_2$CH$_2$CH$_2$CH$_2$CH$_2$CH$_3$	heptane
CH$_3$CH$_2$CH$_2$CH$_2$CH$_2$CH$_2$CH$_2$CH$_3$	octane
CH$_3$CH$_2$CH$_2$CH$_2$CH$_2$CH$_2$CH$_2$CH$_2$CH$_3$	nonane
CH$_3$CH$_2$CH$_2$CH$_2$CH$_2$CH$_2$CH$_2$CH$_2$CH$_2$CH$_3$	decane

Figure 5.2. Alkane nomenclature.

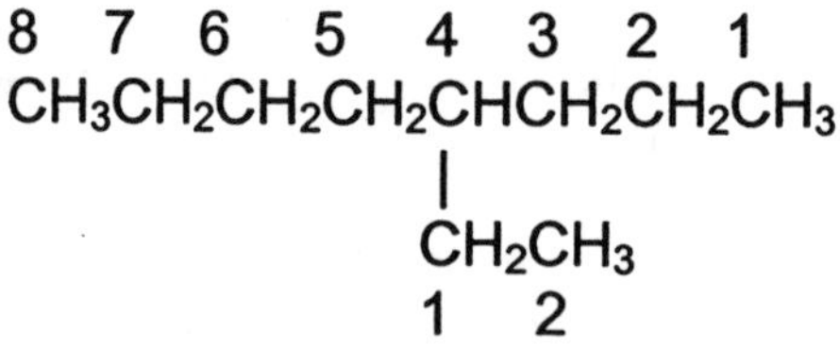

$$\begin{array}{cccccccc} 8 & 7 & 6 & 5 & 4 & 3 & 2 & 1 \end{array}$$
$$CH_3CH_2CH_2CH_2CHCH_2CH_2CH_3$$
$$|$$
$$CH_2CH_3$$
$$\begin{array}{cc} 1 & 2 \end{array}$$

Figure 5.3. 4-ethyloctane.

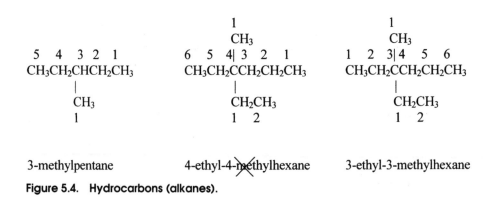

3-methylpentane   4-ethyl-4-~~methyl~~hexane   3-ethyl-3-methylhexane

Figure 5.4. Hydrocarbons (alkanes).

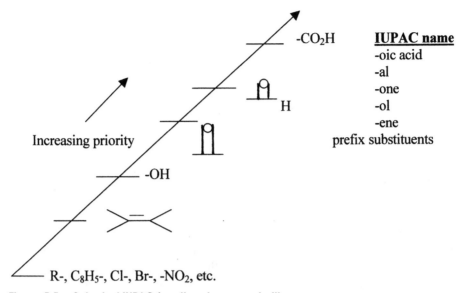

-CO$_2$H   **IUPAC name**
-oic acid
-al
-one
-ol
-ene
prefix substituents

Increasing priority

-OH

R-, C$_8$H$_5$-, Cl-, Br-, -NO$_2$, etc.

Figure 5.5.   Selected IUPAC functional group priorities.

bonds are called alkynes and the IUPAC ending is changed is changed to *-yne* (see figure 5.6).

If four or more carbon atoms are present, the *location* of the double or triple bond(s) must also be denoted. Multiple double or triple bonds are noted by the prefixes *di-* or *tri-*, etc. (see figure 5.7).

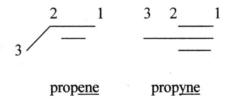

prop**ene**   prop**yne**

Figure 5.6.   Alkenes.

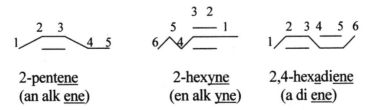

2-pent**ene**   2-hex**yne**   2,4-hex**adiene**
(an alk **ene**)   (en alk **yne**)   (a di **ene**)

Figure 5.7.   Alkynes.

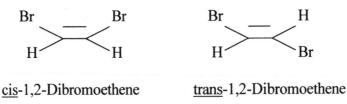

cis-1,2-Dibromoethene   trans-1,2-Dibromoethene

Figure 5.8. Alkenes: Cis/trans.

✔ **Key Point:** Note that an "a" is sometimes inserted before the "di" for ease of pronunciation.

Two groups on the same side of a double bond are called *cis* (from the Latin "on the side"; two groups on opposite sides of a double bond are called *trans*, from the Latin "across" (see figure 5.8). If there are 3 or 4 different atoms attached to the carbons of a double bond, it is impossible to use cis/trans nomenclature (see figure 5.9).

Because of this ambiguity, a more general nomenclature has been developed. In this system, each atom attached to the carbons of a double bond is assigned a priority based on its atomic number (i.e., higher atomic number = higher priority; see figure 5.10). If the groups with the highest priorities are on the *same side* of the double bond, they are designated "Z" (from the German *zusammen,* together). If the groups with the highest priorities are on *opposite sides* of the double bond, they are designated "E" (from the German *entgegan,* across; see figure 5.11).

AROMATIC COMPOUNDS

Just like continuous chain alkanes, a *benzene ring* (aryl group) can be the IUPAC *parent* compound (see figure 5.12). Alkyl substituents or other functional groups are named as prefixes (see figure 5.12). However, if the benzene ring is attached to an alkane containing (1) a functional group or (2) seven or more carbons, then it is relegated to a substituent called a *phenyl* group (see figure 5.13). *Ortho, meta,* and *para* nomenclature refers to substitutions on a benzene ring (see figures 5.14 and 5.15).

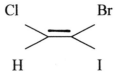

cis or trans?

Figure 5.9. Cis/trans nomenclature?

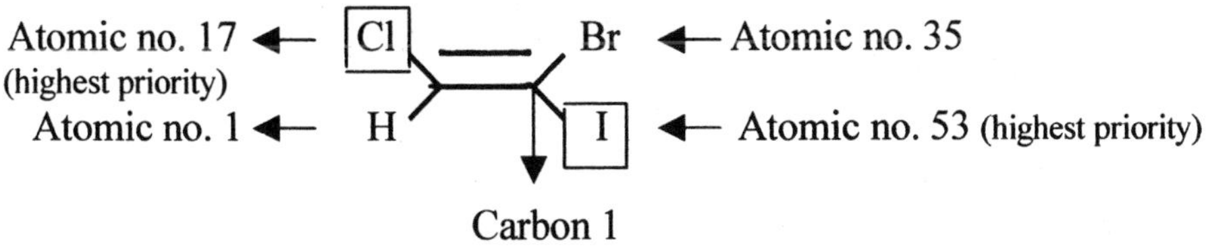

Atomic no. 17 ← Cl   Br ← Atomic no. 35
(highest priority)
Atomic no. 1 ← H   I ← Atomic no. 53 (highest priority)

Carbon 1

(E)-1-Bromo-2-chloro-1-iodoethene
(the highest priority substituents are
on opposite sides of the double bond)

Figure 5.10.   Priority based on atomic number.

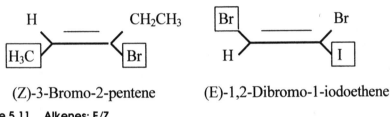

(Z)-3-Bromo-2-pentene   (E)-1,2-Dibromo-1-iodoethene

Figure 5.11.   Alkenes: E/Z.

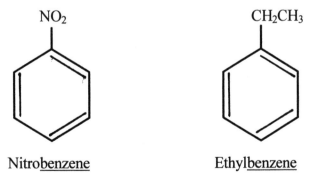

Nitrobenzene   Ethylbenzene

Figure 5.12.   Aromatic compounds.

HALOGENATED COMPOUNDS

Halogenated compounds are named as numbered prefixes on the parent alkane, alkene, alkyne, or benzene rings (see figure 5.16).

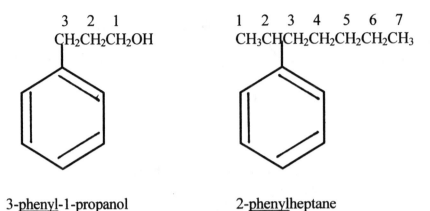

Figure 5.13.   Aromatic compound conflicts.

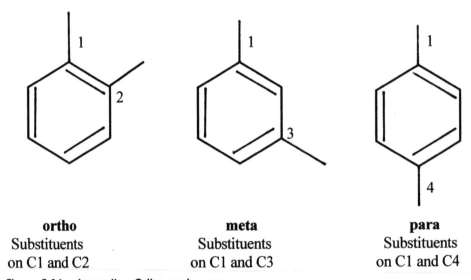

Figure 5.14.   Aromatics: Ortho, meta, para.

ALCOHOLS

IUPAC nomenclature for alcohols is based on the parent hydrocarbon (alkane, alkene, or alkyne), with the final –e changed to –ol. The *hydroxyl group* (-OH) receives the lowest possible prefix number (see figure 5.17).

ALDEHYDES

The IUPAC names for aldehydes always end in –al (see figure 5.18). Note that no numbering is necessary, because the aldehyde group (a carbonyl bonded to an H atom) is always at the end of the carbon chain.

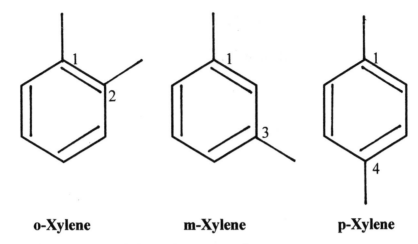

o-Xylene   **m-Xylene**   **p-Xylene**

Figure 5.15.   Aromatics: Ortho, meta, para (cont.).

KETONES

The IUPAC names for ketones end in *–one* (figure 5.18). The *carbonyl* (-CO-) is always given the highest possible number (see figure 5.19).

ORGANIC ACIDS (CARBOXYLIC ACIDS)

The IUPAC name for an organic acid always ends in *–oic acid.* As with aldehydes, no numbering is necessary, because the carboxyl group (-COOH) must be at the end of a carbon chain (see figure 5.20).

ESTERS

The IUPAC name for *esters* end in *–ate.* As with aldehydes and organic acids, no numbering is necessary because the ester functionally must be at the end of a carbon chain (see figure 5.21).

ETHERS AND PEROXIDES (CONFLICTS)

Ethers and *peroxides* are *most commonly* referred to by their trivial names (see figure 5.22). In more complex examples, the prefixes *alkoxy-* and *peroxy-* are used (see figure 5.23).

Chloro (Cl)

Bromo (Br)

Fluoro (F)

Iodo (I)

3   2   1
CH₃CH₂CH₂I

1-Iodopropane

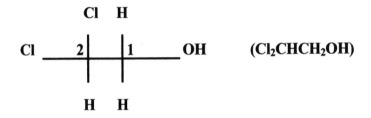

(Cl₂CHCH₂OH)

2,2-Dicloro-1-ethanol

Figure 5.16.   Halogenated compounds.

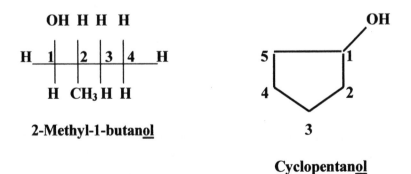

2-Methyl-1-butanol

Cyclopentanol

Figure 5.17.   Alcohols.

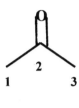

Propanal

3-Methylhexanal

Figure 5.18. Aldehydes.

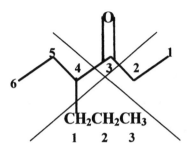

Propanone
(Acetone)

4-Propyl-3-hexanone

Figure 5.19. Ketones.

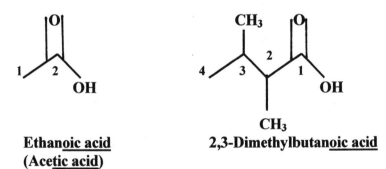

Ethanoic acid
(Acetic acid)

2,3-Dimethylbutanoic acid

Figure 5.20. Organic acids (carboxylic acids).

NITRILES

In IUPAC nomenclature, the number of carbons in the longest chain, including the C of the *nitrile* (CN-), determines the parent (root) alkane, plus the suffix *nitrile* (see figure 5.24). They are also sometimes called *cyano compounds* or *cyanides* (see figure 5.25). Certain *nitriles* are commonly referred to by their organic acid parent, with the *–ic acid* changed to *–nitrile* (or *–onitrile* for ease of pronunciation; see figure 5.26).

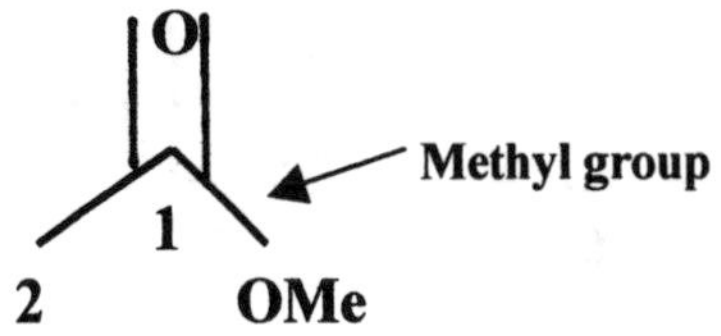

Methyl group

Methyl ethanoate
(Methyl acetate)

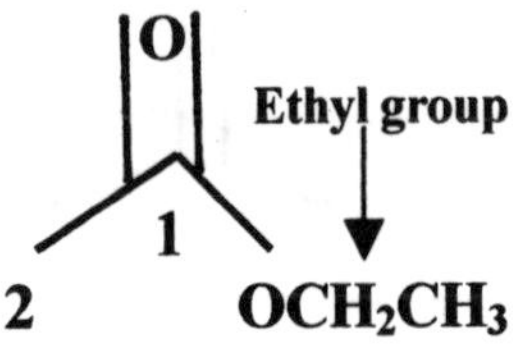

Ethyl group

Ethyl ethanoate
(Ethyl acetate)

Figure 5.21. Esters.

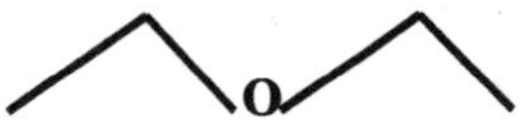

Diethyl ether

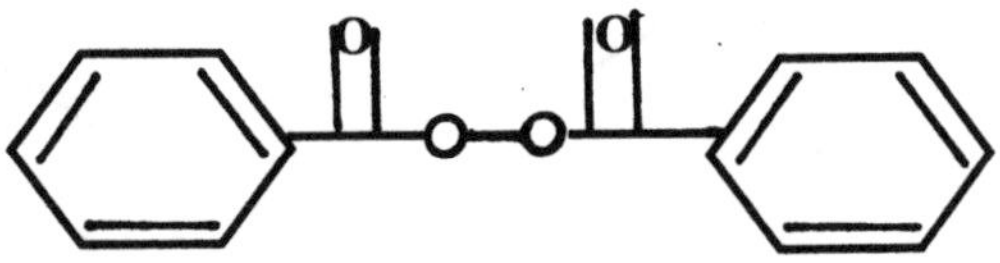

Benzoyl peroxide

$$CH_3CH_2—O—O—CH_2CH_3$$

Diethyl peroxide

Figure 5.22. Ethers and peroxides.

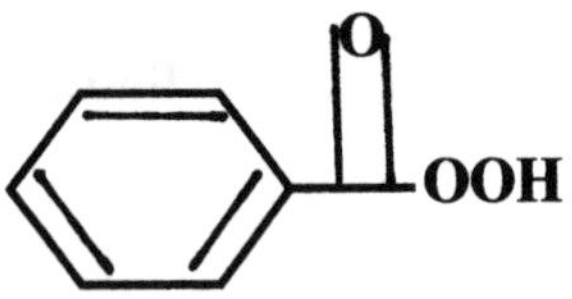

Peroxybenzoic acid

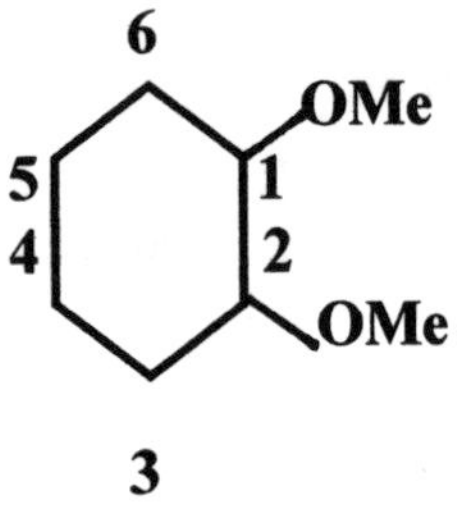

1,2-Dimethoxycyclohexane

Figure 5.23. Ether and peroxide conflicts.

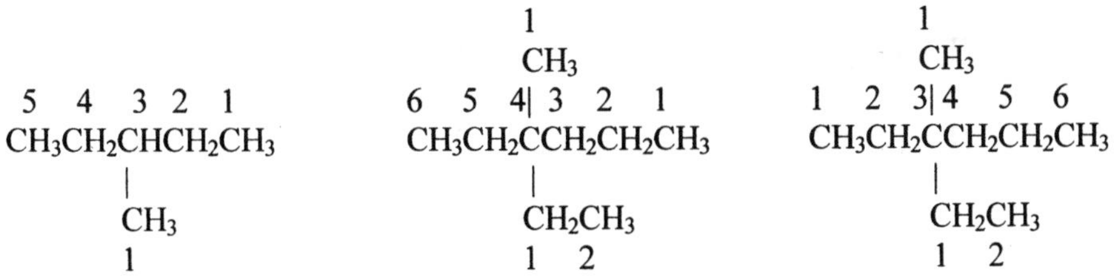

<pre>
                              1   1
                             CH₃   CH₃
  5   4   3 2 1   6   5 4| 3   2   1   1   2 3| 4   5   6
CH₃CH₂CHCH₂CH₃   CH₃CH₂CCH₂CH₂CH₃   CH₃CH₂CCH₂CH₂CH₃
       | | |
      CH₃   CH₂CH₃   CH₂CH₃
       1   1   2   1   2
</pre>

3-methylpentane   4-ethyl-4-methylhexane   3-ethyl-3-methylhexane

Figure 5.24.   Nitriles.

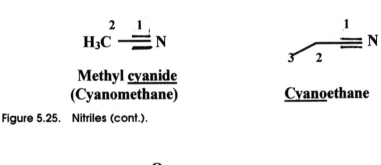

Methyl cyanide
(Cyanomethane)

Cyanoethane

Figure 5.25.   Nitriles (cont.).

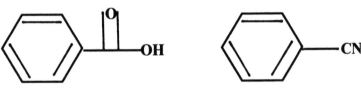

(Benzoic acid)   Benzonitrile

Figure 5.26.   Nitriles (cont.).

AMINES

Amines are simply named for the parent alkane or aryl compound plus the suffix *–amine, -diamine*, etc. (see figure 5.27). In more complex compounds, the prefix *amino-* may be used (see figure 5.28). Lower priority alkyl groups on the N are designated with an *N-alkyl-* prefix (see figure 5.28).

NITRO COMPOUNDS

Nitro compounds are named by the parent alkane, alkene, alkyne, or benzene ring, plus the prefix *nitro-, dinitro*, etc. (see figure 5.29).

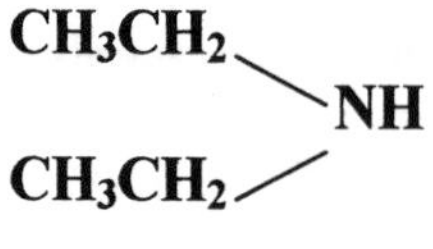

Diethyl<u>amine</u>

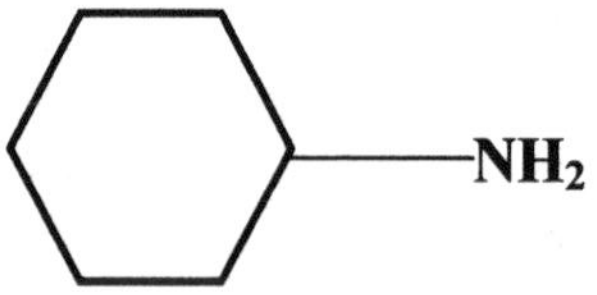

Figure 5.27. Amines.

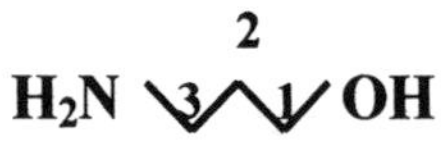

Cyclohexyl<u>amine</u>

3-<u>Amino</u>-1-propanol

Figure 5.28. Amines (cont.).

4-<u>(N-methylamino)</u>butanoic acid

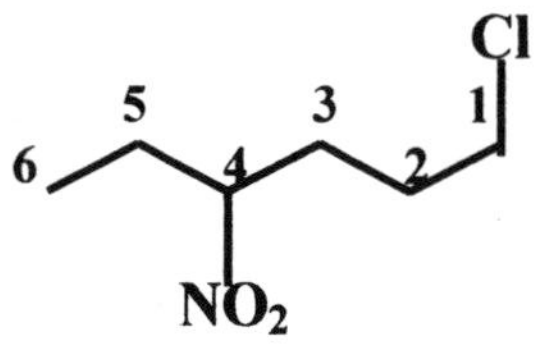

1-Chloro-4-<u>nitro</u>hexane

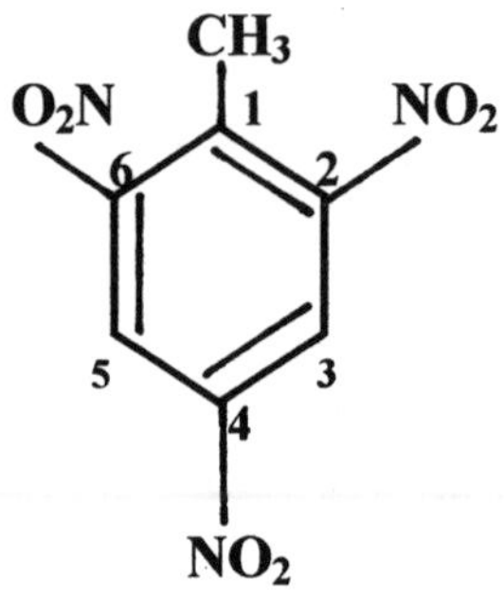

1-Methyl-2,4,6-<u>trinitro</u>benzene

(2,4,6-<u>Trinitro</u>toluene)

Figure 5.29. Nitro compound.

Organic Chemistry

Now that we have reviewed some of the nomenclature used in organic chemistry, we shift our concentration to the basics of organic chemistry itself. Strong, stable bonds between carbon atoms produce complex molecules containing chains and rings. Again, the chemistry of these compounds is called organic chemistry.

Greater than 95% of all known chemicals contain carbon. Carbon is unique because:

- each atom combines with four other atoms;
- it forms chains of atoms; and
- it combines with atoms of other elements.

CARBON BONDS

Carbon forms strong bonds with a variety of nonmetals:

- C-H bond energy = 99 kcal/mol
- C-O bond energy = 84 kcal/mol
- C-Cl bond energy = 79 kcal/mol
- C-N bond energy = 70 kcal/mol

Carbon also forms single and multiple bonds with itself, allowing carbons to be linked together in chains of varying lengths.

- C-C bond energy = 83 kcal/mol
- C=C bond energy = 143 kcal/mol
- C=C bond energy = 196 kcal/mol

✔ **Key Point:** Carbon likes to form bonds so well with itself that it can form multiple bonds to satisfy its valence of four.

In contrast, Si-Si and Ge-Ge bonds are relatively weak.

- Si-Si bond energy = 42 kcal/mol
- Ge-Ge bond energy = 65.4 kcal/mol

Organic compounds primarily contain *covalent bonds* (i.e., equally shared electrons; see chapter 6) between carbon and atoms of hydrogen, oxygen, the halogens, nitrogen, sulfur, and other carbon atoms. The chemical and physical properties of carbon compounds are related to three factors:

- The number of C atoms in the molecule
- The type of bonding in the molecule
- The kind(s) of functional groups in the molecule

NUMBER OF CARBONS

The number of carbon atoms in a molecule has an effect on the molecule's chemical properties: volatility, solubility, mobility, and biodegradability. A standard rule of thumb: Increasing the number of C atoms typically increases:

- Volatility
- Solubility
- Mobility
- Biodegradability
- Boiling point

Examples of increased boiling points with increasing number of carbons are shown in figure 5.30.

TYPE OF BONDING

Organic compounds may contain *single, double,* or *triple covalent bonds* between a single pair of atoms, especially C atoms (see figure 5.31). In general, compounds with double and triple bonds are *more reactive* than compounds with single bonds.

FUNCTIONAL GROUPS

Functional groups play a significant role in determining the reactivity of organic molecules. They behave as a unit in entering or leaving an organic molecule. Certain key functional groups are commonly encountered in environmental investigations.

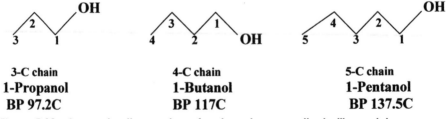

3-C chain	4-C chain	5-C chain
1-Propanol	**1-Butanol**	**1-Pentanol**
BP 97.2C	**BP 117C**	**BP 137.5C**

Figure 5.30. Increasing the number of carbons increases the boiling point.

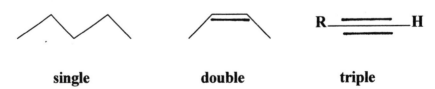

single   **double**   **triple**

Figure 5.31. Type of bonding.

A functional group is an atom or group of atoms that defines the structure of a particular family of organic compounds and determines its properties (see figure 5.32). Stated differently, so-called functional groups determine most of the chemical properties (functions) of organic compounds. Following are the key functional groups:

Methyl, Ethyl	$-CH_3, -CH_2CH_3$
Phenyl	$-C_6H_5$
Chloro	$-Cl$
Bromo	$-Br$
Iodo	$-I$
Fluoro	$-F$
Hydroxyl	$-OH$
Aldehyde	$-COH$
Ketone	$-CO$
Acid	$-COOH$
Ether	$-O-$
Peroxide	$-O-O-$
Ester	$-COO-$
Nitrile	$-CN$
Amino	$-NH_2$
Nitro	$-NO_2$

The functional groups are associated with various *classes* of organic compounds. For example:

• Alcohols (R-OH) contain the hydroxyl (OH) group.
• Amines ($R-NH_2$) contain the amino (NH_2) group.
• Aromatics contain the phenyl (C_6H_5) group.

✔ **Key Point:** The symbol "R" stands for any hydrocarbon radical. A hydrocarbon with one hydrogen removed from it is a radical.

Compounds in the *same class tend to react similarly,* because functional groups play a major role in determining a compound's reactivity. When compounds contain more than one functional group, chemists usually classify the molecule according to the functional group that most dictates its chemical reactivity.

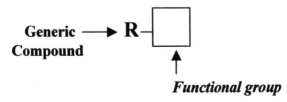

Figure 5.32.   A functional group.

CLASSES OF ORGANIC COMPOUNDS

Hydrocarbons (C, H, only)

Hydrocarbons are compounds containing only hydrogen and carbon. Methyl ($-CH_3$) and ethyl ($-CH_2CH_3$) groups are common hydrocarbon functional groups that also appear in many other organic compounds.

Common hydrocarbons include: methane (CH_4), the chief ingredient of natural gas and propane (C_3H_8); butane (C_4H_{10}), a common ingredient of bottled gas; octane (C_8H_{18}), including isooctane, which is the reference compound for rating gasoline; and paraffin and waxes, which contain longer chains of C.

Because hydrocarbons are common fuels and solvents, they are frequently released to the environment through leaking underground storage tanks and tank overfills. Moreover, since many hydrocarbons are less dense than water, they tend to float on the water table as "free phase hydrocarbons," if released in large quantities.

Aromatics

The phenyl group (i.e., benzene ring) is the most common distinguishing factor for aromatic compounds (see figure 5.33). The term *aromatic* arose from the fact that many of these compounds have distinct odors.

Many aromatic compounds are also classified as hydrocarbons, since they contain only carbon and hydrogen (see figure 5.34). Other common aromatic hydrocarbons include those shown in figure 5.35.

During removal and/or cleanup, the soils surrounding underground storage tanks that contained gasoline are frequently analyzed for *BTEX*: benzene, toluene, ethylbenzene, and xylene. Benzene is regulated as a carcinogen (cancer-causing agent; see figure 5.36). Many other aromatic compounds are also known or suspected carcinogens.

Polychlorinated biphenyls (PCBs) are halogenated organics that are very persistent in the environment (see figure 5.37). They are synthetic chemical compounds consisting of chlorine, carbon, and hydrogen. PCBs belong to a family of organic compounds known as chlorinated hydrocarbons and can be found as a clear to yellow oily liquid or waxy solid. The 1976 Toxic Substance Control Act (TSCA) prohibited any further manufacture of PCBs in the United States.

When people or animals ingest PCBs, they are stored in the fatty tissue and then slowly released into the blood stream. Even at low exposure levels in the environment, the concentration of PCBs in fatty tissue can accumulate to a high level. This process is called *bioaccumulation*. The PCBs accumulate in the fatty tissue of organisms low

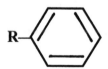

Figure 5.33. Aromatics (phenyl).

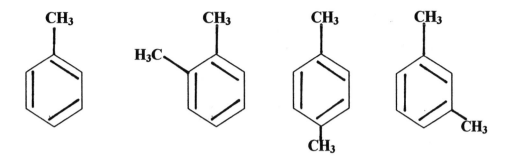

CH_3
CH_3
H_3C
CH_3
CH_3
CH_3
CH_3
Toluene
Xylenes

Figure 5.34. Aromatics (common hydrocarbon solvents).

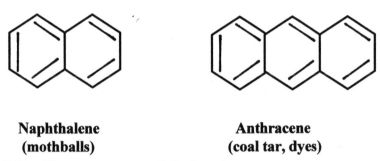

Naphthalene
(mothballs)
Anthracene
(coal tar, dyes)

Figure 5.35. Other common aromatic hydrocarbons.

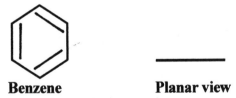

Benzene
Planar view

Figure 5.36. Aromatics (phenyl carcinogens).

in the food chain and are magnified when consumed by the animals in the higher food chain. This process is termed *biomagnification.* As PCBs bioaccumulate in organisms and biomagnify in the food chain, they create health hazards at all levels. The short-term health hazards for people include irritation to the eyes, nose, and throat. Serious health effects are often not immediately apparent, but can develop and persist for months or years, depending on the exposure. PCBs can cause liver damage, reproductive problems, a severe acne-like rash (chloracne), and damage to the nervous system.

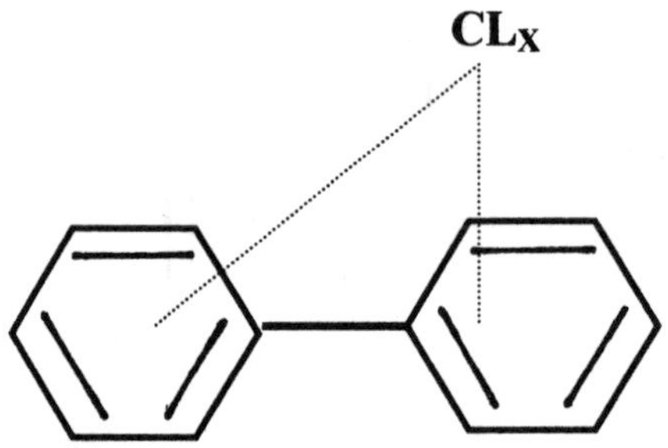

$$\mathbf{CL_X}$$

Figure 5.37. Aromatics (PCBs).

The primary application of PCBs nationwide has been as a dielectric (insulating) fluid for high-power electrical equipment. Typically, large institutions purchase power at a reduced cost by receiving the power at high voltage and then dropping down the voltage with their own transformers. Other equipment used for high-power applications includes large capacitors, voltage regulators, and switches—all of which may contain PCB liquid insulating material. PCBs may also be found in the capacitors and "potting material" of fluorescent light ballasts. Light ballasts that contain PCBs are regulated for disposal under federal PCB regulations; particularly, if the ballasts are leaking, they must be disposed of as TSCA waste in an approved incinerator. It is recommended that, when more than a handful of nonleaking ballasts are being disposed of, they be taken to a TSCA landfill or approved recycling facility. This will help protect the environment from any future leaks as well as eliminate your liability for future PCB contamination.

DDT, which has been banned in the United States, also persists and biomagnifies in the food chain (see figures 5.38 and 5.39).

✔ **Key Point:** According to USEPA (EPA-540/1-75-022, 1975), certain characteristics of DDT that contributed to its early popularity, particularly its persistence, later became the basis for public concern over possible hazards involved in its use. Although scientists voiced warnings against such hazards as early as the mid-1940s, it was Rachel Carson's *Silent Spring* in 1962 that stimulated widespread public concern over use of DDT. After Carson's alert to the public concerning the dangers of improper pesticide use and the need for better pesticide controls, it was only natural that DDT, as one of the most widely used pesticides of the time, should come under intensive investigation.

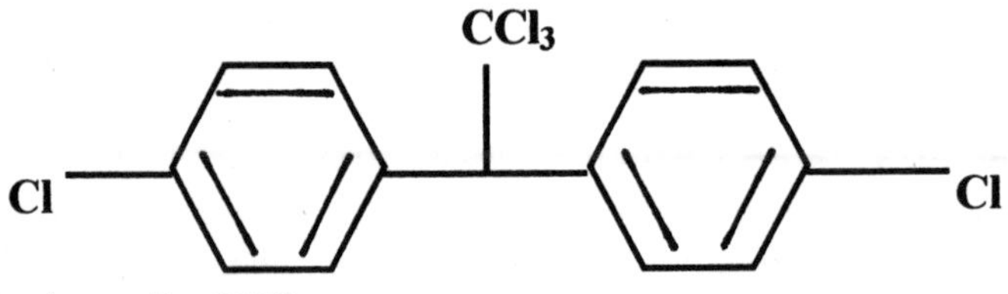

$$\mathbf{CCl_3}$$

Cl ——   —— Cl

Figure 5.38. Aromatics (DDT).

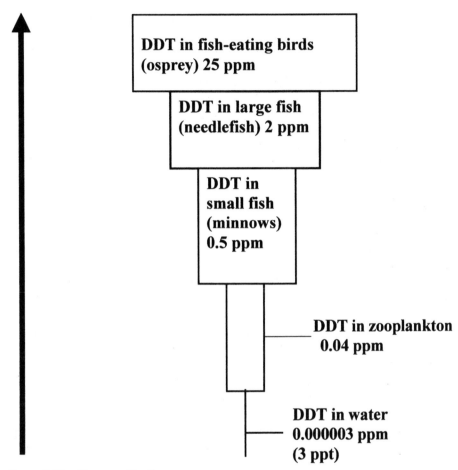

Concentration Increase – 10 million Times

DDT in fish-eating birds (osprey) 25 ppm

DDT in large fish (needlefish) 2 ppm

DDT in small fish (minnows) 0.5 ppm

DDT in zooplankton 0.04 ppm

DDT in water 0.000003 ppm (3 ppt)

Figure 5.39.   Biomagnification.

The 411 on DDT: The Irony of Trying to Control Nature

Since publication of the work of Rachel Carson, we have changed our view on the efficacy of DDT for mosquito control. In thinking that we could shape nature and create the perfect disease-free world, we allowed nature to counter our interruption and shape its own environment. By regarding DDT as foolproof—the ultimate mosquito control and the cure-all to end all malaria-related problems—we seriously miscalculated. DDT was not foolproof. Blinded to the realities of the situation, we touted our victories and failed to notice that nature was fighting back. Because of our ignorance, the lives of millions around the world were ultimately, and still are, being shaped by nature. As Rachel Carson pointed out, nature has its own ways of controlling insects, and by our widespread assault on nature we decreased the environment's own ability to control them and increased the negative affects of what we sought to eliminate.

Dioxins result from pesticide manufacture and PCB incineration (see figure 5.40).

✔ **Key Point:** Dioxin is a highly toxic industrial by-product of industrial processes involving chlorine. Sources of dioxin include paper and pulp mills, hazardous waste incineration, sludge from waste facilities, cement kilns that burn chemical waste, and the manufacturing of PVC plastics and some pesticides. The human defects include learning disabilities, infertility, suppressed immune functions, and reduced IQs and hyperactive behavior in children.

Halogens (Cl, Br, I, F)

The elements of Group VII of the periodic table are known as the halogens, or "salt formers." The halogens are too reactive to occur free in nature, but their compounds are widely distributed. Halogens consist of chlorine, bromine, iodine, fluorine, and astatine. (Astatine [At] is so rare that it is of no importance to environmental chemistry.) In the chemically combined state, the halogens are very abundant and widely distributed.

Halogenated compounds are of concern in environmental chemistry because the compounds they form are typically both poisonous and persistent.

✔ **Key Point:** *Persistent* compounds do not biodegrade easily in the natural environment and therefore tend to bioaccumulate (e.g., PCBs).

All the halogens form diatomic molecules (e.g., Cl_2, Br_2). All halogens have sharp, disagreeable odors and attack the skin and mucous membranes of the nose and throat. Common halogenated compounds include:

Hydrogen halides (HF, HBr, HCl, HI) are gases at room temperature and are extremely soluble in water, forming very strong "inorganic" acids, such as hydrochloric acid.

Carbon tetrachloride (CCl_4), methylene chloride (CH_2Cl_2), and perchloroethylene

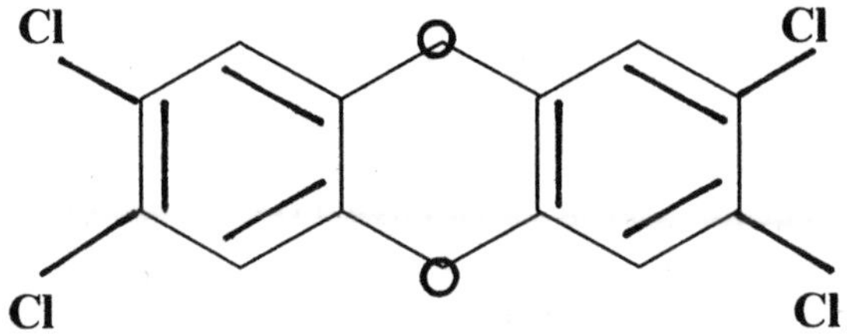

Figure 5.40. Aromatics (dioxins).

(perc, C_2Cl_4). Since many of these solvents are very dense and are not very soluble in water, they can form dense, nonaqueous phase liquids (DNAPLs).

Household *bleaches* such as Clorox and Purex are approximately 5% solutions of sodium hypochlorite NaOCl (which forms hypochlorous acid [HOCl] in water):

$$NaOCl + H_2 \rightarrow HOCl + NaOH$$

Chlorofluorocarbons (CFCs) are also known by the trade name Freon. CFCs are commonly used as refrigerants and were once used as propellants. These compounds were thought to be especially stable and nontoxic. Widely manufactured CFCs include CFC-11, CFC-12, CFC-113, and others.

Halons are related compounds that contain bromine and were once widely used for fire suppression until their role in depleting the ozone layer became apparent. Halon 1301 is one of the most widely used.

✔ **Key Point:** Both chlorofluorocarbons (CFCs) and halons are being phased out because of their role in destroying stratospheric ozone (O_3).

The destruction of ozone is a complicated, multistep process. The production of CFCs and halons was curtailed in the United States beginning in 1989, in accordance with the 1986 Montreal Protocol on Substances that Deplete the Ozone Layer. The most likely substitutes are hydrogen-containing chlorofluorocarbons (HCFCs) and hydrogen-containing fluorocarbons (HFCs).

Alcohols (-OH)

A hydrocarbon that has a hydroxyl (-OH) group (a functional group) is called an alcohol. All alcohols can be recognized by the *-ol* ending in their names.

The simplest alcohol is *methanol* (methyl alcohol), which is also known as wood alcohol because it can be obtained from wood. Wood alcohol is extremely poisonous. Breathing its vapors or drinking it may cause blindness or death.

CH_3OH (methyl alcohol)

Ethanol (C_2H_5OH), called ethyl alcohol, grain alcohol, or simply alcohol, is the most important of the alcohols. Ethanol is found in alcoholic beverages and is an important solvent and precursor for the production of other chemicals.

CH_3CH_2OH (ethyl alcohol)

Phenol (C_6H_5OH) is an aromatic alcohol that is an important reactant in the production of plastics (see figure 5.41).

Ethylene glycol ($C_2H_6O_2$) is a *diol* (i.e., contains 2 OH groups; see figure 5.42) and is used as a solvent and as an antifreeze in automobile radiators.

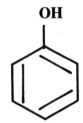

Figure 5.41. Alcohols (-OH).

Figure 5.42. Alcohols (diol).

✔ **Key Points:** Lower-molecular-weight alcohols tend to be both *volatile* and *readily biodegradable*. More complex alcohols and aromatic alcohols may be more persistent in the environment.

Aldehydes (-COH)

Alcohols represent the first stage of oxidation of hydrocarbons. Further oxidation produces aldehydes, compounds containing the group CHO (see figure 5.43).

Formaldehyde (HCOH), the simplest aldehyde (see figure 5.44), is a colorless gas with a pungent and irritating odor; it is poisonous and used as a germicidal agent and preservative. Dissolved in water (formalin) it is used as a germicide, preservative, and embalming agent. Formaldehyde is used in large quantities in the production of plastics and adhesives.

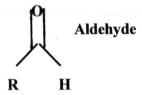

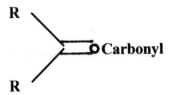

Figure 5.43. Aldehydes (-COH).

Figure 5.44. Aldehydes (formaldehyde).

✔ **Key Point:** The formaldehyde found in plastics, glues, coloring agents, plywood, and manmade fibers, etc. has been implicated in contributing to *indoor air quality* problems. As a result, many manufacturers have significantly decreased or entirely phased out the use of formaldehyde in their products.

Acetaldehyde (CH_3CHO), a liquid, and other aldehydes produce the flavor and scent of many alcohols. Acetaldehyde is an intermediate in the manufacture of acetic acid (see figure 5.45).

✔ **Key Point:** Aldehydes tend to be both *volatile* and *reactive* and therefore typically do not persist in the environment without undergoing further reactions.

Ketones (C = O)

Ketones, like aldehydes, contain the carbonyl group, $C=O$. (See figure 5.46.) These carbonyl groups are anywhere in a chain except in the end, and the carbonyl carbon is bonded only to other carbons. The simplest and most important ketone is *acetone*, which is an important solvent. *Methyl ethyl ketone* (MEK) is another important solvent (see figure 5.47).

Many ketones have strong aromas that are used for perfumes and other scents. The simpler ketones tend to be relatively volatile and therefore typically do not play a major role in environmental investigations of soil and groundwater.

Organic Acids (-COOH)

If aldehyde continues to be oxidized, its molecules will gain additional oxygen. Eventually, an acid will form. The -COOH is the functional group that identifies all

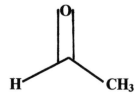

Figure 5.45. Aldehydes (acetaldehyde).

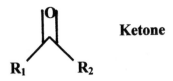

Ketone

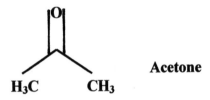

Carbonyl

Figure 5.46. Ketones (C=O).

Acetone

MEK

Figure 5.47. Ketones (acetone and MEK).

organic acids. All organic acids, known as carboxylic acids, are weak because they are only partially ionized. One of the simplest acids is *acetic acid,* vinegar (i.e., vinegar is a 5% solution of acetic acid in water). *Citric acid* ($C_6H_8O_7$) is commonly found in citrus fruit; it is a tribasic acid with three carboxylic acid groups (see figure 5.48).

Hydrocarbon → alcohol → aldehyde or ketone → acid

✔ **Key Point:** Inorganic acids such as HCl and HBr are much stronger acids than organic acids such as acetic acid. Organic acids are typically referred to as "weak" acids, since they do not ionize to a large extent in an aqueous solution.

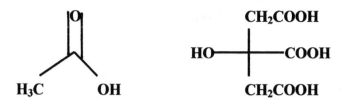

acetic acid   **citric acid**

Figure 5.48. Organic acids (-COOH).

Esters (-COOR)

When acids react with alcohols, a class of compounds called esters is formed. The whole range of carboxylic acids and alcohols can be reacted to form esters. Like ketones, esters tend to have pleasant odors and flavors, which are used in cosmetics, beverages, confections, and medicines. Starch, proteins, and rubber are natural polymers. Probably the best-known esters are amyl acetate (banana oil) and methyl salicylate (oil of wintergreen). Polyester, used primarily in clothing, is a polymer containing repetitive ester units.

Ethers (-O-) and Peroxides (-O-O-)

Ethers are compounds obtained from alcohols by the elimination of a molecule of water from two molecules of the alcohol. (See figure 5.49.) Ethers contain the characteristic linkage -C-O-C-. They are typically extremely volatile, and their ability to form *peroxides* makes them extremely dangerous (see figure 5.50).

Hydrogen peroxide (HOOH) is the simplest peroxide. *Hyperperoxides* (ROOH) and *dialkl peroxides* (R_1OOR_2) are notorious explosion hazards because they can be extremely sensitive to shock, heat, and/or friction. The following is a standard classification for common peroxide-forming chemicals:

Class A—chemicals that form explosive levels of peroxides without concentration or evaporation.

Class B—chemicals that form explosive levels of peroxides on concentration or evaporation.

Class C—chemicals that may auto-polymerize as a result of peroxide accumulation.

Nitriles (CN)

Two commonly used nitriles are *acetonitrile* (CH_3CN) and *acrylonitrile* (CH_2CHCN). Acetonitrile is widely used as a precursor in the chemical industry. Acrylonitrile is highly reactive and metabolizes to release *cyanide* (HCN).

Common Ethers

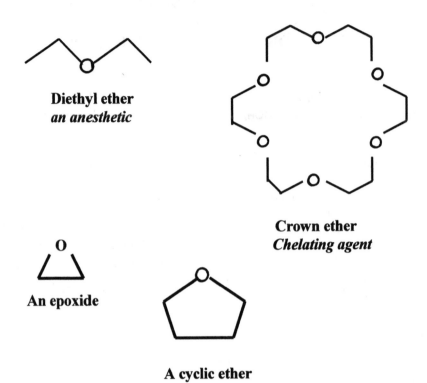

Diethyl ether
an anesthetic

Crown ether
Chelating agent

An epoxide

A cyclic ether

Figure 5.49.   Ethers (-O-) and peroxides (-O-O-).

$$R_1\text{-}O\text{-}R_2 \xrightarrow{\ O_2\ } R_1\text{-}O\text{-}O\text{-}R_2$$

Figure 5.50.   Ether formed into peroxide.

Amines (-NH₂)

Organic derivatives of ammonia are called amines. The simpler amines are rapidly taken into the body, where they react with water to raise the pH of the tissue to harmful levels. An example of a simple amine is *methylamine* (CH_3NH_2), which is a colorless, highly flammable gas with a strong odor; it has the characteristic odor of rotten fish (see figure 5.51).

Some of the *aromatic amines* have been shown to be *carcinogens.*

✔ Interesting Point: Nylon is a *polyamine.*

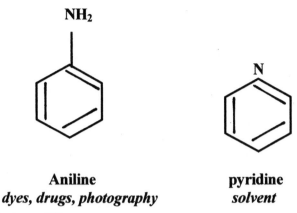

Aniline
dyes, drugs, photography

pyridine
solvent

Figure 5.51. Amines (-NH$_2$).

Nitro Compounds (-NO$_2$)

Trinitrotoluene (TNT) is a widely used explosive (see figure 5.52). *Nitrogen dioxide* (NO$_2$) and *nitric oxide* (NO) are important air pollutants that are collectively referred to as *NO$_x$*. Atmospheric reactions convert NO$_x$ to *nitric acid,* a contributor to acid rain.

Additional Reading

Brown, H. B., and C. Foote. *Organic Chemistry.* Philadelphia, Pa.: Saunders College Publishing, 1997.

McMurry, S. *Organic Chemistry.* 5th ed. Pacific Grove, Calif.: Brooks/Cole, 1999.

Wade, L.G., Jr. *Organic Chemistry.* 5th ed. Upper Saddle River, N.J.: Prentice-Hall, 2002.

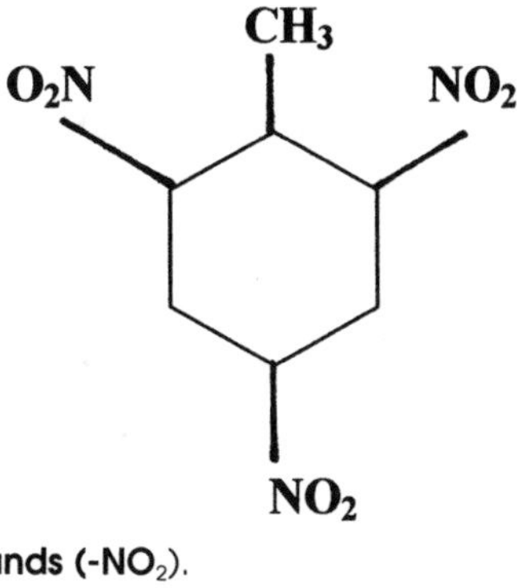

Figure 5.52. Nitro compounds (-NO$_2$).

Summary

Strong, stable bonds between carbon atoms produce complex molecules containing chains and rings. The chemistry of these compounds is called *organic chemistry*.

I began this chapter by explaining that the key to learning organic chemistry is recognizing and becoming familiar with its nomenclature. The International Union of Pure and Applied Chemistry (IUPAC) provides the applicable nomenclature. Just as important to gaining an understanding of organic chemistry is to recognize and understand the chemistry of the important functional groups; see figure 5.53.

Generally, then, organic chemistry is the study of *carbon* compounds.

- Carbon is different from all other elements because it can covalently bond to itself in many chains and designs.
- There are more different carbon compounds than all noncarbon compounds put together.
- Carbon forms covalent bonds with itself and other atoms.
- Carbon may form single, double, or triple bonds with itself.
- Carbon cannot form four covalent bonds with itself.
- Hydrocarbons are organic compounds composed only of carbon and hydrogen.
- There are two major groups of hydrocarbons: the aliphatics and the aromatics.
- Aromatic hydrocarbons contain ring structures.
- Aliphatic hydrocarbons include all organic compounds that do not possess a benzene ring; they are not aromatic.
- The alkanes are saturated hydrocarbons that contain only single bonds; they are

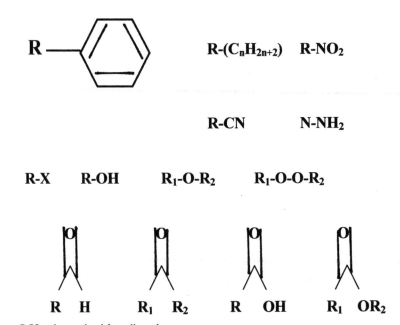

Figure 5.53.   Important functional groups.

better known as the methane series. The *-ane* at the end of the names of all alkane compounds identifies them.

- Isomers are compounds with the same chemical formula but different arrangements of atoms.
- Alkenes are hydrocarbons that contain one or more carbon-carbon double bonds.
- Alkynes contain one or more carbon-carbon triple bonds.
- The functional group for all alcohols is -OH.
- A functional group is a group of atoms in a molecule that is responsible for the properties of the compound. For example, the -OH is the functional group of an alcohol.
- Alcohols can be oxidized to aldehydes or ketones.
- Carbohydrates are aldehydes and ketones.
- Esters are produced by the reaction of acids with alcohol.
- -O- is the functional group for ethers.
- -O-O- is the functional group for peroxides.
- $-NO_2$ is the functional group for nitro compounds.

New Word Review

Alcohol—an organic compound that has the hydroxyl (-OH) group as its functional group.

Aldehyde—compounds containing the group -CHO.

Aliphatic hydrocarbons—all organic compounds that do not possess a benzene ring.

Alkanes—aliphatic hydrocarbons in which there are only single bonds between carbon atoms.

Alkenes—aliphatic hydrocarbons in which there is at least one double bond between carbon atoms.

Alkynes—aliphatic hydrocarbons that have a triple bond between carbon atoms.

Aromatic hydrocarbons—cyclic or ring compounds of carbon that have special properties.

Ester—a compound produced by the reaction of acids with alcohols.

Hydrocarbons—compounds of carbon and hydrogen.

Ketones—compounds that include the carbonyl group.

Isomers—compounds with the same formula but different molecular structure or atomic arrangement.

Organic chemistry—the study of carbon compounds.

Chapter Review Questions

MATCHING

Column A	Column B
5.1. C_nH_{2n}	1. 3 pairs of electrons
5.2. C_6H_6	2. Pentane

5.3. -C=C-   3. Pentyne
5.4. -R-O-R-   4. Alkane
5.5. aliphatic   5. Pentene
5.6. C_5H_{10} 6. Ether
5.7. C_5H_8 7. Straight-chain hydrocarbon
5.8. C_5H_{12} 8. Alcohol
5.9. -R-OH   9. Aromatic hydrocarbon
5.10. $C_nH_{(2n + 2)}$ 10. Alkene

5.11. All the following compounds are alkanes except:
    a. C_7H_{16}
    b. $C_{12}H_{26}$
    c. C_5H_{12}
    d. C_3H_6
5.12. What is the total number of carbon atoms in a molecule of pentyne?
    a. 2
    b. 1
    c. 5
    d. 3
5.13. Ethane, ethene, and ethyne are all similar in that they are:
    a. hydrocarbons
    b. aromatic
    c. all unsaturated
    d. all saturated
5.14. The formula for a saturated hydrocarbon is _____.
5.15. Benzene is an example of a(n) _____ hydrocarbon.
5.16. The general formula for an alcohol is:
    a. R-H
    b. R-COOH
    c. R-H
    d. R-COOR
5.17. The functional group for an organic acid is called a(n) _____.
5.18. The functional group for an aldehyde is _____.
5.19. ROH is the general formula for any _____.
5.20. Pentanol is a(n) _____-carbon alcohol.

CHAPTER 6

Chemical and Physical Reactions

One of the important tasks of chemistry is to study how substances can be identified or distinguished from each other, that is, a study of properties. Two important chemical properties that need to be studied are chemical and physical reactions. Such studies are also essential in determining how substances can be used in human endeavors.

Topics in This Chapter

- Types of Chemical Reactions
- Reactions Rates (Kinetics)
- Types of Physical Reactions
- Examples of Physical vs. Chemical Changes

Fundamental to any study of chemistry are *chemical reactions*, *reaction rates*, and *physical reactions*. However, before discussing chemical reactivity, it is important to review the basics of *electron distribution* and chemical and physical changes, and to discuss the important role heat plays in both chemical and physical reactions.

ELECTRON DISTRIBUTION

Simply, electron distribution around the nucleus is key to understanding chemical reactivity. Only electrons are involved in chemical change; the nuclei of atoms are not altered in any way during chemical reactions. Electrons are arranged around the nucleus in a definite pattern or series of "shells."

In general, only the "outer shell" or "valence" electrons (i.e., the ones farthest from the nucleus) are affected during chemical change. The *valence number* or *valence* of an element indicates the number of electrons involved in forming a compound and the number of electrons it tends to gain or lose when combining with other elements.

Positive valence indicates giving up electrons. Negative valence indicates accepting electrons.

TYPES OF BONDING

If these valence electrons are *shared* with other atoms, then a *covalent bond* is formed when the compound is produced. If these valence electrons are *donated* to other atoms, then an *ionic bond* is formed when the compound is produced.

If an atom gains or loses one or more valence electrons, it becomes an *ion* (charged particle). *Cations* are positively charged particles. *Anions* are negatively charged particles.

PHYSICAL AND CHEMICAL CHANGES

You will recall that a chemical change is the change physical substances undergo when they become new or different substances. To identify a *chemical change*, you look for observable signs, such as color change, light production, smoke, bubbling or fizzing, and presence of heat. Simply, in a chemical change, the matter cannot easily go back to its original state.

A *physical change* occurs when objects undergo a change that does not change their chemical nature. A physical change involves a change in physical properties. Physical properties can be observed without changing the type of matter. Examples of physical properties include texture, size, shape, color, odor, mass, volume, density, and weight. Simply, a physical change is a change you can observe by using your five senses.

HEAT AND CHEMICAL/PHYSICAL REACTIONS

An *endothermic reaction* is a chemical reaction that absorbs energy and where the energy content of the products is more than that of the reactants; heat is taken in by the system. An *exothermic reaction* is a chemical reaction that gives out energy and where the energy content of the products is less than that of the reactants; heat is given out from the system.

Types of Chemical Reactions

Chemical reactions are of fundamental importance throughout chemistry and related technologies. Although experienced chemists can sometimes predict the reactions that will occur in a new chemical system, they may overlook some alternatives. Further, they are usually unable to make reliable predictions when the chemistry is unfamiliar to them.

✔ **Key Point:** There are few tools available to assist in predicting chemical reactions, and none at all for predicting the novel reactions that are of greatest interest. (For more information on predicting chemical reactions, see K. K. Irikura and R. D. Johnson III, in Additional Reading.)

There are so many chemical reactions (the Merck Index, for example, lists 425 "named" reactions; see figure 6.1 for two of them) that it is helpful to classify them into general types. Four general types of chemical reaction are:

• Combination
• Decomposition
• Replacement
• Double replacement

COMBINATION REACTIONS

In a combination (or synthesis) reaction, two or more simple substances combine to form a more complex compound. For example, copper (an element) + oxygen (an element) = copper oxide (a compound).

✔ **Key Point:** There are many pairs of reactants that combine to give a single product. The reactions happen when it is energetically favorable to do so.

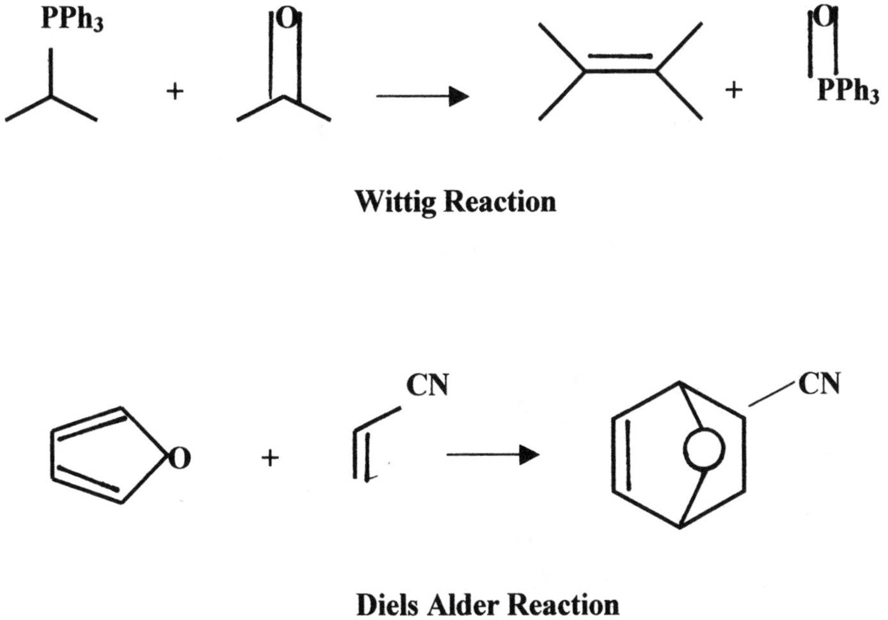

Wittig Reaction

Diels Alder Reaction

Figure 6.1. "Named" reactions.

DECOMPOSITION REACTIONS

Decomposition reactions occur when one compound breaks down (decomposes) into two or more substances or its elements. Basically, combination and decomposition reactions are opposites.

$$\text{Hydrogen peroxide} \rightarrow H_2O + O_2$$

REPLACEMENT REACTIONS

Replacement reactions involve the substitution of one uncombined element for another in a compound. Two reactants yield two products.

$$\text{Fe (iron)} + H_2SO_4 \text{ (sulfuric acid)} = H_2 \text{ (hydrogen)} = FeSO_4 \text{ (iron sulfate)}$$

DOUBLE REPLACEMENT

In a double replacement reaction, parts of two compounds exchange places to form two new compounds. Two reactants yield two products.

$$\text{sodium hydroxide} + \text{acetic acid} = \text{sodium acetate} + \text{water}$$

✔ **Key Point:** Note that these four classifications of chemical reactions are not based on the type of bonding, which generally can be either covalent or ionic, depending on the reactants involved.

✔ **Interesting Point:** According to the law of conservation of mass, no mass is added or removed in chemical reactions.

SPECIFIC TYPES OF CHEMICAL REACTIONS

Specific types of chemical reactions include:

- Hydrolysis and neutralization
- Oxidation/reduction (redox)
- Chelation
- Free radical reaction
- Photolysis and polymerization
- Catalysis
- Biochemical reaction and biodegradation

Hydrolysis ("hydro" means water and "lysis" means to break) is a decomposition reaction involving the splitting of water into its ions and the formation of a weak acid or base or both. *Neutralization* is a double replacement reaction that unites the H⁻ ion and an acid with the OH⁻ ion of a base, forming water and a salt.

$$\text{acid} + \text{acid} \rightarrow \text{salt} + \text{water}$$

For example

$$\text{sodium hydroxide} + \text{acetic acid} = \text{sodium acetate} + \text{water}$$

Oxidation/reduction (redox) reactions are combination reactions, replacement, or double replacement reactions that involve the gain and loss of electrons (i.e., changes in valence). Oxidation and reduction always occur simultaneously, so that one reacting species is oxidized while the other is reduced.

> ✔ **Key Point:** When an atom, either free or in a molecule or ion, loses electrons, it is *oxidized*, and its oxidation number increases. When an atom, either free or in a molecule or ion, gains electrons, it is *reduced*, and its oxidation number decreases.

Chelation is a combination reaction in which a *ligand* (such as a solvent molecule or simple ion) forms more than one bond to a central ion, giving rise to complex ions or coordination compounds (see figure 6.2).

> ✔ **Interesting Point:** The term *chelate* was first applied in 1920 by Sir Gilbert T. Morgan and H. D. K. Drew (*J. Chem. Soc.*, 1920, 117, 1456), who stated: "The adjective chelate, derived from the great claw or chela (chely—Greek) of the lobster or other crustaceans, is suggested for the caliperlike groups which function as two associating units and fasten to the central atom so as to produce heterocyclic rings."

Free radical reactions involve any species with an unpaired electron. Free radical reactions frequently occur in gas phase, often proceed by chain reaction, are often initiated

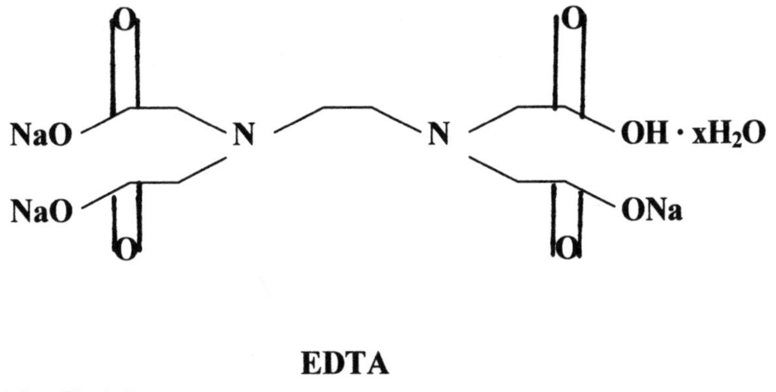

EDTA

Figure 6.2. Chelation.

by light, heat, or reagents that contain unpaired electrons (such as oxides or peroxide decomposition products) that are often very reactive.

✔ **Interesting Point:** A common free radical reaction in aqueous solution is electron transfer, especially to the hydroxyl radical and to ozone.

Photolysis (*photo* meaning "light" and *lysis* meaning "to break") is generally a decomposition reaction in which the adsorption of light produces a photochemical reaction. Photolysis reactions form free radicals that can undergo other reactions. For example, the chlorine molecule can dissociate in the presence of high-energy light (e.g., UV light).

$$Cl_2 + UV \text{ energy} \rightarrow Cl + Cl$$

✔ **Important Point:** Photolysis is an important nonthermal technology used for treating dioxin and furan hazardous wastes.

Polymerization is a combination reaction in which small organic molecules are linked together to form long chains (see figure 6.3) or complex two- and three-dimensional networks. Polymerization occurs only in molecules having double or triple bonds and is usually dependent on temperature, pressure, and a suitable catalyst.

✔ **Key Point:** *Catalysts* are agents that change the speed of a chemical reaction without affecting the yield or undergoing permanent chemical change.

Free radical "chain reactions" are a common mode of polymerization. The term *chain reaction* is used because each reaction produces another reactive species (i.e., another free radical) to continue the process (see figure 6.4).

✔ **Important Point:** Plastics are perhaps the most common polymers, but there are also many important biopolymers such as polysaccharides.

Biochemical reactions are reactions that occur in living organisms. *Biodegradation* is a decomposition reaction that occurs in microorganisms to create smaller, less complex inorganic and organic molecules. Usually the products of biodegradation are molecular forms that tend to occur in nature.

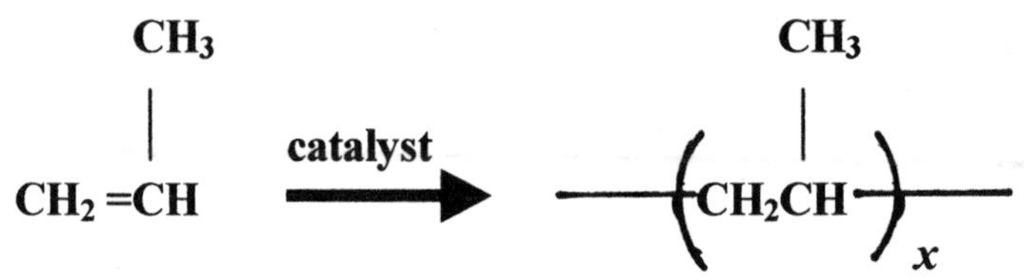

Figure 6.3. Polymerization.

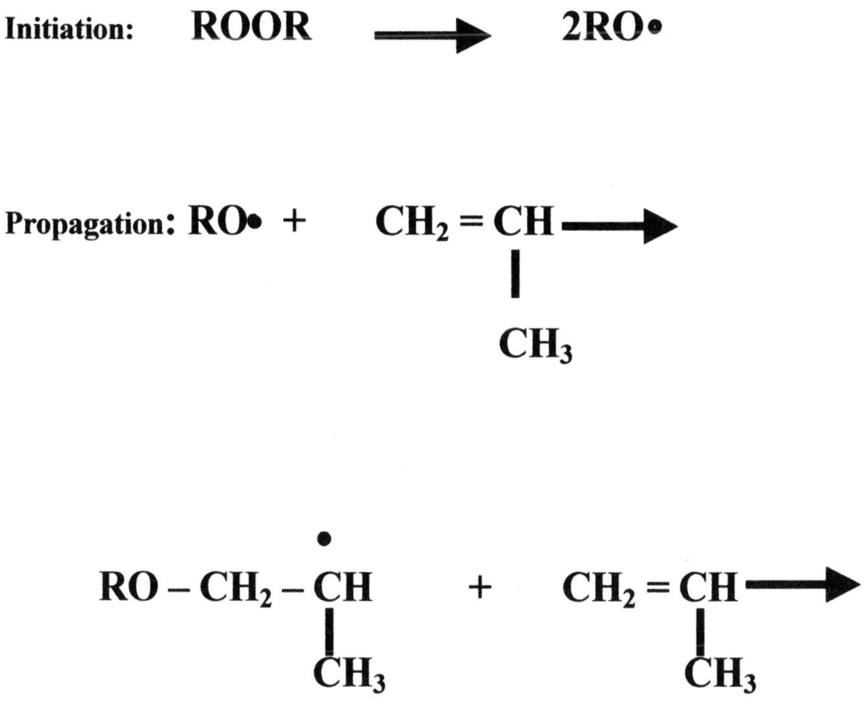

Initiation: ROOR $\longrightarrow$ 2RO•

Propagation: RO• + $CH_2 = CH \longrightarrow$

$$CH_3$$

$$RO - CH_2 - \overset{\bullet}{CH} \quad + \quad CH_2 = CH \longrightarrow$$
$$\underset{CH_3}{|} \qquad\qquad\qquad \underset{CH_3}{|}$$

Figure 6.4.   Chain reactions.

Reaction Rates (Kinetics)

The *rate* of a chemical reaction is a measure of how fast the reaction proceeds, that is, how fast reactants are consumed and products are formed.

$$A + B \text{ (reactants)} \rightarrow C + D \text{ (products)}$$

The rate of a given reaction depends on many variables (factors), including temperature, the concentration of the reactants, catalysts, the structure of the reactants, and the pressure of gaseous reactants or products. Without considering extreme conditions, reaction rates generally increase with:

Increasing temperature. Two molecules will react only if they have enough energy. By heating the mixture, the energy levels of the molecules involved in the reaction are raised. Raising the temperature means the molecules move faster (the kinetic theory).

Increasing concentration of reactants. Increasing the concentration of the reactants increases the frequency of collisions between the two reactants (collision theory).

Introducing catalysts. Catalysts speed up chemical reactions by lowering the activation energy. Only very small quantities of a catalyst are required to produce a dramatic change in the rate of the reaction. This is so because the reaction proceeds by a different pathway when the catalyst is present. Interestingly, adding more catalyst will make absolutely no difference.

Increasing surface area. The larger the surface area of a solid, the faster the reaction will be. Smaller particles have a bigger surface area than larger particles for the same mass of solid. There is a simple way to visualize this, called the Bread-and-butter Theory. If we take a loaf of bread and cut it into slices, we get an extra surface onto which we can spread butter. The thinner we cut the slices, the more slices we get and the more butter we can spread on them. By chewing our food we increase the surface area so that digestion goes faster.

Increasing the pressure on a gas to increase the frequency of collisions between them. By increasing pressure, molecules are squeezed together so that the frequency of collisions between them is increased.

Reaction rates are also affected by each reaction's *activation energy,* the energy the reactants must reach before they can react (see figure 6.5).

✔ **Key Point:** A catalyst may be recovered unaltered at the end of the reaction.

Both "forward" and "backward" reactions can occur, each with a different reaction rate and associated activation energy.

$$A + C \rightleftarrows C + D$$

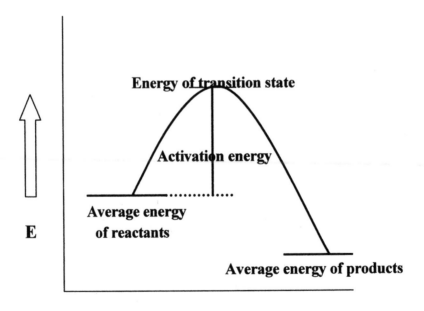

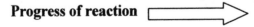

Figure 6.5. Activation energy.

For example, in a dissociation reaction that occurs readily, initially the dissociation takes place at a faster rate than recombination. Eventually, as the concentration of dissociated ions builds up, the rate of recombination catches up with the rate of dissociation. When the forward and backward reactions eventually occur at the same rate, a *state of equilibrium* is reached. The apparent effect is no change, even though both the forward and backward reactions are still occurring.

✔ **Key Point:** Note that the point at which equilibrium is reached is *not fixed* but is dependent on such variables as temperature, reactant concentration, pressure, reactant structure, etc.

Types of Physical Reactions

Knowledge of the physical behavior of wastes and hazardous wastes has been used to develop various unit processes for waste treatment that are based on physical reactions. These operations include the following:

- Phase separation
- Phase transition
- Phase transfer

Phase separation involves separation of components of a mixture that is already in two different phases. Types of phase separation include *filtration, settling, decanting,* and *centrifugation.*

   Phase transition is a physical reaction in which a material changes from one physical phase to another. Types of phase transition include *distillation, evaporation, precipitation,* and *freeze drying* (lyophilization).

   Phase transfer consists of the transfer of a solute in a mixture from one phase to another. Two examples of phase transfer include *extraction* and *sorption* (i.e., transfer of a substance from a solution to a solid phase).

Examples of Physical vs. Chemical Changes

Table 6.1 illustrates everyday examples of physical versus chemical changes.

Additional Reading

Irikura, K. K., and R. D. Johnson III. "Predicting Unexpected Chemical Reactions by Isopotential Searching," *Journal of Physical Chemistry* A, no. 104 (February 2000): 2191–94.
Missen, R. W., et al. *Introduction to Chemical Reactions Engineering and Kinetics.* New York: John Wiley & Sons, 1999.
Oxlade, C. *Materials Changes & Reactions (Chemicals in Action).* Heinemann Library, 2002.

Table 6.1. Physical vs. chemical changes.

Physical Changes	Chemical Changes
Tearing or cutting paper	Burning matches
Melting and freezing wax	Burning gas
Breaking grass	Frying an egg
Crushing rocks	Baking soda and vinegar
Dissolving sugar or powdered drink mix	Burning steel wool and a candle

Summary

- During an endothermic reaction, heat is taken in by the system.
- During an exothermic reaction, heat is given out from the system.
- Four general types of chemical reactions include combination, decomposition, replacement, and double replacement.
- Specific types of chemical reactions include hydrolysis and neutralization, oxidation/reduction, chelation, free radical reaction, photolysis and polymerization, catalysis, and biochemical reaction and biodegradation.
- The rate of a given reaction depends on many variables, including temperature, the concentration of the reactants, catalyst, the surface area of the reactants, and the pressure of gaseous reactants or products.
- Physical reactions include phase separation, phase transition, and phase transfer.

New Word Review

Biodegradation—a decomposition reaction that occurs in microorganisms to create smaller, less complex inorganic and organic molecules.

Catalysts—substances that change the speed of a chemical reaction without affecting the yield or undergoing permanent chemical change.

Chelation—a reaction that involves a ligand that is attached to a central ion by bonds from two or more donor atoms.

Combination reaction—a reaction of two or more substances to give another substance.

Decomposition reaction—a reaction in which one compound breaks down into two or more substances.

Double replacement reaction—a reaction in which the parts of two compounds exchange places to form two new compounds.

Free radical reaction—a reaction that involves any species with an unpaired electron.

Hydrolysis—a decomposition reaction involving the splitting of water into its ions and the formation of a weak acid or base or both.

Neutralization—a double replacement reaction that unites the H$^-$ ion and an acid with the OH$^-$ ion of a base, forming water and a salt.

Oxidation/reduction—a reaction in which oxidation numbers change as electrons are lost by one atom and gained by another.

Phase separation—the separation of components of a mixture that is already in two different phases.

Phase transfer—consists of the transfer of a solute in a mixture from one phase to another.

Phase transition—a physical reaction in which a material changes from one physical phase to another.

Photolysis—a decomposition reaction in which the adsorption of light produces a photochemical reaction.

Polymerization—a combination reaction in which small organic molecules are linked together to form long chains.

Reaction rate—a measure of how fast the reaction proceeds.

Replacement reaction—reaction in which one uncombined element is substituted for another compound.

Chapter Review Questions

6.1. During a chemical reaction, _____ react(s) to form _____.
   a. a reactant, a product
   b. products, a reactant

6.2. The law of conservation of _____ states that no mass is added or removed in chemical reactions.

6.3. When two substances combine to form a third substance, this type of reaction is called a _____ reaction.

6.4. An upward pointing arrow indicates that a _____ is being produced during a chemical equation.

6.5. Increasing the surface area _____ the rate of chemical reactions.

6.6. Higher temperature means molecules have _____ energy leading to _____ reactions.

6.7. A _____ lowers the activation energy required for the reaction.

6.8. The _____ that form during chemical reaction are called the precipitate.

6.9. a + b → c + energy is an example of a(n) _____ reaction.

6.10. Compound → element + element represents a _____ type of chemical reaction.

6.11. Exothermic reactions give off _____.

6.12. Higher concentrations mean _____ reactions between molecules for faster reactions.

6.13. Temperature affects the _____ of a reaction.

6.14. Catalysts can be used over and over, they _____ get used up during a chemical reaction.

6.15. The more collisions in a reaction, the _____ the reaction will occur.

6.16. Endothermic reactions require _____.

6.17. Increased exposed surface areas should _____ reaction rates.

6.18. In a _____ reaction, two elements combine to form a compound.

6.19. A + B → AB is an example of a _____ reaction.

6.20. AB + CD → CB + AD is an example of a _____ reaction.

CHAPTER 7

Understanding Material Safety Data Sheets (MSDSs)

Incidents have occurred when sodium hypochlorite solutions have been accidentally mixed with incompatible materials resulting in the release of elemental chlorine gas and/or a violent reaction. People have been seriously injured as a result of these events.

Since more and more wastewater treatment plants are now using or considering using sodium hypochlorite solutions (often referred to as bleach) as a disinfectant or treatment chemical, it is important to know what potential hazards with this chemical must be addressed. [The fact that] it comes as a liquid (aqueous solution) does not mean that the release of chlorine gas cannot occur. Steps must be taken to prevent this from happening in your plant or facility.

—The Chlorine Institute (www.CL2.com)

Topics in This Chapter

- Material Safety Data Sheets (MSDSs)
- Chemical Measurements
- Chemical Compatibility

For the EH&S profession no OSHA standard has had more profound influence on the health and safety of workers in the workplace than has 29 CFR 1910.1200, the Hazard Communication Standard.

HAZARD COMMUNICATION STANDARD

Why is the Hazard Communication Standard, or HazCom, so important to the EH&S practitioner? OSHA has estimated that more than 32 million workers are exposed to 650,000 hazardous chemical products in more than 3 million American workplaces. This poses a serious problem for exposed workers and their employers.[1]

OSHA points out that the basic goal of a hazard communication program is to be sure employers and employees know about work hazards and how to protect them-

selves. Along with the Clean Air and Clean Water Acts and other important safety/
health and environmental regulations, a hazard communication program should help
reduce the incidence of chemical-source illness and injuries.[2]

HazCom (29 CFR 1910.1200) is designed "to ensure that the hazards of all
chemicals produced or imported are evaluated, and that information concerning their
hazards is transmitted to employers and employees."

Material Safety Data Sheets (MSDSs) are just one aspect of these regulations,
which also include requirements for employee training, labeling, chemical inventories,
and other forms of warning. In this chapter, my focus is on MSDSs, because they are
at the heart of the standard and because they are important tools that the EH&S
professional simply cannot be without. Moreover, along with all EH&S practitioners,
employees who work with, around, or near listed hazardous materials (HAZMATs)
must have a thorough understanding of MSDSs, their uses, and their limitations.
Because MSDS parameters are often given in common units used in analytical chemistry, we also provide a brief discussion of *chemical measurements: unit analysis* and
conversion factors.

✔ **Important Point:** The Hazard Communication Program is just one of several
   EH&S regulatory requirements in which chemistry is directly represented and
   applied. Knowledge of the physical and chemical properties of substances is fundamental to managing chemicals safely.

Material Safety Data Sheets (MSDSs)

The requirement that manufacturers and distributors provide MSDSs to their customers became effective on November 25, 1985. The regulation does not specify the exact
format for an MSDS; however, OSHA has provided a *suggested* format.[3]

Although OSHA does not dictate the format for an MSDS exactly, it does impose
minimum requirements regarding the information manufacturers and importers must
include in the document. At a minimum, manufacturers and importers must provide,
in English,[4] the following information on all MSDSs:
• Identity of the hazardous chemical as it appears on the chemical's label
• Physical and chemical characteristics of the hazardous chemical
• Physical hazards of the hazardous chemical
• Health hazards of the hazardous chemical
• Primary route(s) of entry
• Exposure limits, including the OSHA permissible exposure limit, ACGIH Thresh-
  old Limit Value, and any other exposure limit used or recommended by the chemi-
  cal manufacturer, importer, or employer who is preparing the MSDS
• OSHA's (and other recognized authority's) finding that the chemical is a potential
  carcinogen, if applicable
• Precautions regarding safe handling and use
• Control measures
• Emergency and first aid procedures

- Date of the MSDS preparation or its last change
- Name, address, and telephone number of the manufacturer, importer, employer, or other responsible party who prepared or distributed the MSDS[5]

✔ **Key Point:** The Chemical Manufacturers Association (CMA) has developed a *voluntary* standard, published in 1993 as ANSI Z400.1-1993, The American National Standard for Hazardous Industrial Chemicals—MSDS Preparation.

ANSI MSDS FORMAT (RECOMMENDATIONS ONLY)

The American National Standards Institute (ANSI) format has sixteen sections:

Section 1: Chemical Product and Company Identification, including its chemical and common names (for example, brand name = Clorox, chemical name = sodium hypochlorite, common name = bleach)

Section 2: Composition, Information on Ingredients (even in parts as small as 1%)

Section 3: Hazards Identification (stability, reactivity, flammability, explosive, corrosive, etc.)

Section 4: First Aid Measures

Section 5: Fire Fighting Measures

Section 6: Accidental Release Measures

Section 7: Handling and Storage

Section 8: Exposure Controls, Personal Protection

Section 9: Physical and Chemical Properties

Section 10: Stability and Reactivity

Section 11: Toxicological Information

Section 12: Ecological Information

Section 13: Disposal Considerations

Section 14: Transport Information

Section 15: Regulatory Information

Section 16: Other Information

In the following, I describe only those sections *specifically related* to the chemical properties of the substance.

Section 1: Chemical Product and Company Identification

The MSDS is required to clearly identify the chemical substance. This can be accomplished by identification of its chemical name, trade name(s), and/or synonyms. In addition, the manufacturer/distributor's mailing address and telephone number is required. Whatever name is used to identify the substance, it must relate the MSDS to the substance label and shipping documents.

Section 2: Composition, Information on Ingredients

All hazardous components of the substance (nonhazardous ingredients are listed separately), the Chemical Abstract Service (CAS) number(s), Permissible Exposure Level

(PEL), Threshold Limit Values (health effects data), and trade secrets disclaimer must be listed. Additionally, the *Percent by Weight* (%) or *Percent Composition by Weight* is also provided in this section. For example, a particular grade of benzene may be composed of 99.8% benzene and 0.2% toluene.

Section 3: Hazards Identification

Provides a description of the material's appearance, color, odor, etc. Health, physical, and environmental hazard data that may be important to emergency response personnel is also provided.

Section 5: Fire Fighting Measures

Describes the fire and explosive properties of the material. This section also identifies appropriate extinguishing media and provides firefighting instructions.

✔ **Key Point:** Any fire requires fuel, oxygen, and an ignition source.

Fire fighting parameters are also provided, such as flash point, flammable limits (i.e., Upper Explosion Limit [UEL] and Lower Explosive Limit [LEL]), and the autoignition temperature. Below the LEL, there is an insufficient fuel concentration to ignite the mixture. Above the UEL, there is an insufficient concentration of oxygen to ignite the mixture.

✔ **Key Point:** LEL is sometimes listed as LFL (lower flammability level) and UEL is listed as UFL (upper flammability limit)—the terms are interchangeable.

✔ **Interesting Point:** When trying to think of LEL and its counterpart, UEL, it helps to use an example that most people are familiar with—the combustion process that occurs in the automobile engine. When an automobile engine has a gas/air mixture that is below the LEL, the engine will not start because the mixture is too lean. When the same engine has a gas/air mixture that is above the UEL, it will not start because the mixture is too rich (engine is flooded). However, when the gas/air mixture is anywhere between the LEL and UEL levels, the engine should start (see figure 7.1).

Another important parameter is *flash point*. Flash point is the temperature at which a liquid flashes (i.e., ignites) when exposed to an ignition source. The lower the flash point, the greater the fire hazards.

Flammability Limits (UFL and LFL) indicate the range in which a compound is a potential fire hazard (i.e., can ignite when exposed to an ignition source).

Autoignition temperature is the lowest temperature at which a flammable liquid–gas mixture will ignite without spark or flame.

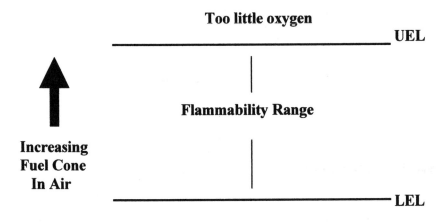

Figure 7.1.   UEL/LEL range.

Section 9: Physical and Chemical Properties

Physical data describe the material's appearance, odor, physical state (solid, liquid, and gas). Pay close attention to the *physical description* of a chemical to help confirm (1) the identity of the material and (2) its age and/or purity.

✔ **Caution:** Never smell a chemical to determine its identity!

Chemical data include:
- Corrosivity (pH)
- Vapor pressure and vapor density
- Boiling and melting points, sublimation data
- Solubility/miscibility
- Density and specific gravity

If applicable, the MSDS will report the pH of the material as an indication of *corrosivity* (see figure 7.2).

Vapor pressure is the pressure exerted when a solid or liquid is in equilibrium with its own vapor (see figure 7.3). Vapor pressure (reported in mm Hg) is a function of

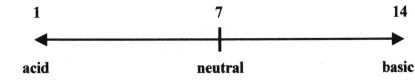

Figure 7.2.   The pH scale.

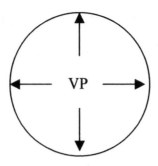

Figure 7.3. Vapor pressure.

the compound and the temperature (e.g., rubbing alcohol [isopropanol], 33 mm Hg @ 20°; benzene, 74.6 mm Hg @ 20°C).

✔ **Rule of Thumb:** Higher vapor pressure materials generally evaporate more quickly.

MSDSs also provide physical and chemical data about the state (solid, liquid, gas) of the material and the temperature(s) at which it changes states:
* Boiling points
* Melting points
* Sublimation temperatures

A *phase transition* is the transformation of a compound from solid to liquid, from solid to vapor, or from liquid to vapor (or vice versa; see figure 7.4).

Boiling point is a measure of the temperature at which a liquid undergoes a phase transition from a liquid to a vapor. Boiling point is typically given for compounds that are liquid at room temperature. This number is useful as an indication of volatility.

✔ **Key Point:** Note that boiling points decrease with decreasing pressure (see figure 7.5).

Melting point (MP) is a measure of the temperature at which a solid undergoes a phase transition to a liquid. Melting points are typically provided for compounds that are

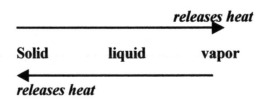

Figure 7.4. Phase transition.

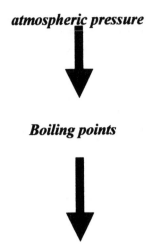

atmospheric pressure

Boiling points

Figure 7.5. Boiling points.

solids at room temperature. This number may be useful when determining appropriate storage containers and/or containment.

Sublimation is a change in state from solid to gas. This number may be important in determining how to appropriately store a waste, for example.

Solubility is a measure of a compound's ability to dissolve in another liquid, especially water. Solubility is reported in grams per volume (typically g/100ml H_2O) at a specified temperature (usually 25°C unless otherwise stated).

✔ **Key Point:** Solubility is often reported as "solubility in water." Solubility may also be reported as a percentage. For example, benzene is only slightly soluble in water (0.1% to 1%). The general solubility rule of thumb: *Like dissolves like.*

Miscible is synonymous with the term *soluble* with reference to liquids.

✔ **Key Point:** If solubility data are reported as "miscible" or "100% miscible," it means that the liquid is completely soluble in the solvent (typically water) in any ratio.

Density is the mass of a unit volume of a substance. It is typically reported in g/ml. *Specific gravity* is the ratio of the density of a liquid or solid to that of water (at a specified temperature). A value less than 1 indicates that the compound is less dense than water. Specific gravity data are important in determining whether a compound will sink or float within the water column (assuming it is not soluble in water).

Vapor density is a measure of the density of a compound's vapors. It is typically reported relative to the density of ambient air, which is arbitrarily assigned a density of 1. A value of <1 indicates that the compound's vapors are less dense than air; that is, they will rise within the atmosphere. Vapor density >1 will sink in air, seeking the low point.

✔ **Key Point:** Vapor density data are important in determining whether or not the compound will remain in the *breathing zone* (i.e., between a person's shoulders and top of head), resulting in possible inhalation hazards and /or *oxygen deficiency* (i.e., displacement of oxygen), especially in confined spaces.

Section 10: Stability and Reactivity

- Instability under specific conditions. For example, for the nerve agent VX the stability is described as follows: Relatively stable at room temperature. Unstabilized VX of 95% purity decomposes at a rate of 5% a month at 71°C.
- Violent polymerization
- Violent decomposition
- Incompatibility
- Other hazardous properties

Section 11: Toxicological Information

Toxicology data describe specific exposure limits that lead to acute (i.e., short-term) health effects, such as eye and skin irritation, as well as intermediate and long-term (i.e., chronic) effects of exposure. *Lachrymator* (under Eye Irritation) means that the substance causes a burning sensation in the eyes that produces tears. LD_{50} (Lethal Dose 50) is the dosage via various routes (through skin absorption or ingestion) that results in death for one half of the animal test population. For example, the oral LD_{50} for benzene is 3.8 g/kg. LC_{50} (Lethal Concentration 50) is the vapor concentration of a chemical at which one half of the animal test population dies upon inhalation.

✔ **Key Point:** It is difficult to measure the "dose" of an inhaled chemical, so scientists measure the concentration in the air to determine an LC_{50}.

A *carcinogen* is a substance that can cause cancer. A *suspected carcinogen* is a potential chemical carcinogen for which limited, variable, or inconclusive data is available. A *mutagen* is a substance that can cause changes in DNA. A *teratogen* is a substance that can cause birth defects. And *IDLH* stands for Immediately Dangerous to Life and Health.

✔ **Key Point:** The IDLH concentration is the airborne concentration that poses a threat of death, immediate or delayed permanent adverse health effects, and/or inability to escape from the contaminated environment in the event of respiratory PPE failure.

Other sources of MSDS data include the Internet, CD-ROMs, and chemical and toxicology reference books.

✔ **Key Point:** The MSDS is a good starting point for chemical and physical data that will assist in the safe handling of a chemical. It is important to remember, however, that the MSDS could be incomplete or out of date. Although valuable as sources of generic names and chemicals, the toxicity information in an MSDS should be confirmed by other sources..

Chemical Measurements

In the MSDS, either the Permissible Exposure Limits (PEL) or Threshold Limit Values (TLV) of the hazardous components of the substance are listed. Therefore, to correctly apply the data provided in the MSDS, the user practitioner must be familiar with common units used in analytical chemistry and be able to convert to/from ppm and ppb and make other unit conversions. For a list of prefixes used see table 7.1.

COMMON UNITS IN ANALYTICAL CHEMISTRY

Grams (g):

milligram (mg)	10^{-3} g
micrograms (μg)	10^{-6} g (μ = mu for micrograms)
nanograms (ng)	10^{-9} g
kilograms (kg)	10^{3} g

Liters (L or l):

Millimeters (ml)	10^{-3} l

Table 7.1. Using prefixes as multipliers.

Prefix	Symbol	Factor
exa	E	10^{18}
peta	P	10^{15}
tera	T	10^{10}
giga	G	10^{9}
mega	M	10^{6}
kilo	k	10^{3}
hecto	h	10^{2}
deca	da	10^{1}
deci	d	10^{-1}
centi	c	10^{-1}
milli	m	10^{-3}
micro	μ	10^{-6}
nano	n	10^{-9}
pico	p	10^{-12}
femto	f	10^{-15}
atto	a	10^{-16}

COMMON UNITS IN ANALYTICAL CHEMISTRY

Parts per million (ppm): one part per million parts
Parts per billion (ppb): one part per billion parts

✔ **Key Point:** When considering ppm, think of a full shot glass of water at the bottom of a standard-sized swimming pool—full shot glass of water $=$ 1 ppm relative to total pool water.

✔ **Key Point:** Milligrams per liter (mg/l) is a unit expressing the concentration of chemical constituents in solution as weight (milligrams) of solute per unit volume (liter) of water. One thousand micrograms per liter is equivalent to one milligram per liter. For concentration less than 7,000 mg/l, the numerical value is the same as for concentrations in parts per million.
mass of substance

$$\text{ppm} = \frac{\text{mass of substance}}{\text{mass of sample}}$$

CONVERTING TO PARTS PER MILLION AND PARTS PER BILLION

Why does mg/l $=$ ppm? 1 liter of water is defined as equal to a weight of 10^6 mg. Therefore, mg/l $=$ mg/10^6 mg *or* 1 mg per 1 million mg *or* 1 part per million parts.

CONVERTING TO PARTS PER MILLION AND PARTS PER BILLION

Why does mg/kg $=$ ppm? mg/kg $= 10^{-3}$ g/10^3 g *or* $(10^{-3}$ g/10^3 g) $(10^3/10^3)$.

✔ **Important Point:** Remember that multiplying the numerator (top) and the denominator (bottom) of a fraction by the same number does not change its value because you are multiplying by 1.

mg/kg $= 10^{-3}$ g/10^3 g *or* $(10^{-3}$ g/10^3 g) $(10^3/10^3)$ *or* 10^0 g/10^6 g.

✔ **Key Point:** Remember that to multiply exponents, add them. Any number raised to the zero power (i.e., having an exponent of zero) equals 1.

mg/kg $= 10^{-3}$ g/10^3 g *or* $(10^{-3}$ g/10^3 g) $(10^3/10^3)$ *or* 10^0 g/10^6 g *or* 1 g/10^6 g.

✔ **Key Point:** Remember that 10^6 equals one million.

mg/kg $= 10^{-3}$ g/10^3 g *or* $(10^{-3}$ g/10^3 g) $(10^3/10^3)$ *or* 10^0 g/10^6 g *or* 1 g/10^6 g *or* 1 gram per million grams *or* 1 part per million parts.

OTHER UNIT CONVERSIONS (UNIT ANALYSIS)

To convert from one unit to another, follow these three simple steps:

1. Look up the equivalence between the units (e.g., 3 feet = 1 yard).
2. Set up the calculation so that units you are trying to obtain are "on top"; i.e., don't cancel out.
3. Perform the calculation, being sure that the correct units cancel.

Here is another example of conversion: How do you convert 15 centimeters into meters? We know that 1 meter equals 100 centimeters.

$$1 \text{ m} = 100 \text{ cm}$$

We therefore set up the equation to put meters "on top" (because we want meters).

$$(15 \text{ cm}) (1 \text{ meter}/100\text{cm}) = ?$$

Perform the calculation, being sure the correct units cancel.

$$(15 \text{ cm}) (1 \text{ m}/100\text{cm}) = 0.15 \text{ meters}$$

✔ **Key Points:** When performing unit conversions, always perform a *unit analysis* to be sure that you have calculated the conversion correctly.

Chemical Compatibility

We opened this chapter describing a safety and environmental problem that can occur when incompatible substances are accidentally mixed. That narrative points to the need for the proper management of chemicals, which begins with proper storage and inventory control. Obviously properly storing hazardous materials and hazardous wastes is critical to chemical safety and plays a large part in the overall responsibility of the EH&S professional.

The basic principles for managing hazardous materials include (1) protecting containers from physical damage; (2) isolating flammables from heat, flames, or sparks; and (3) isolating incompatible materials. Common examples of chemical incompatibility:

- Acids react violently with bases or water.
- Phenolics react with acids, oxidizers, or hypochlorites.
- Cyanides generate toxic fumes when exposed to acids.
- Flammable solvents are incompatible with heat, fire, or explosives.

✔ **Key Point:** Note that some wastes that are not chemically incompatible still often need to be segregated, simply because mixing increases the cost and/or difficulty of disposing of the chemicals.

There are three basic approaches available for managing chemical incompatibility: (1) check MSDS information; (2) when in doubt, isolate; and (3) perform compatibility analysis. Many different systems have been developed to determine chemical compatibility. The USEPA Hazardous Waste Compatibility Protocol uses:

- 41 reactivity group numbers (RGNs)
- 9 reactivity codes
- A list of RGNs for chemical substances

REACTIVITY GROUPS

The first 34 reactivity groups are chemically similar materials:

- Acids (3)
- Metals (4)
- Hydrocarbons (3)
- Alkalis or bases

The remaining seven reactivity groups (101–107) have similar hazardous properties:

- Combustible/flammable
- Explosive
- Polymerizing
- Oxidizing/reducing
- Water and mixtures containing water
- Water reactive

REACTIVITY CODES

Temperature:

- Heat generation (H)
- Fire (F)

Gas generation (often with heat):

- Innocuous and non-flammable gas generation (G)
- Flammable gas generation (GF)

Other reactions:

- Explosion (E) (often with heat)
- Violent polymerization (P)
- Solubilization of toxic substances (S)

Unknown ("catchall"):

May be hazardous but unknown (when in doubt, isolate)

LIST OF CHEMICAL SUBSTANCES

Common substances found in hazardous waste streams are listed in 3 columns:

- Reactivity Group Number
- Chemical/trade name (marked with *) in alphabetical order
- Synonyms or common names

✔ **Key Point:** When storing *any* amount of different chemicals in close proximity, be sure to check their chemical compatibility—when in doubt, ISOLATE, ISO-LATE, ISOLATE.

Summary

All workers are required by law to have access to Material Safety Data Sheets, and all manufacturers and importers must provide MSDSs for hazardous (or potentially hazardous) chemicals.

An MSDS is a document that provides information about the chemical or product. It may vary in style and content, but it must contain certain required information. State and federal law requires that all manufacturers and distributors of chemical products provide the end user with a manufacturer-specific MSDS. The primary method of accessing MSDSs is through electronic databases.

The goal of the MSDS is to provide the user with a summarized, multisource resource with basic information regarding a chemical's properties and related effects, personnel protective equipment (PPE) necessary to protect the user, first aid treatment in the event of an exposure, how to respond to accidents, and the planning that may be needed in order to safely handle a spill or daily operations involving it.

Though an important source of health and safety information, the MSDS should not be used exclusively to evaluate chemical hazards, but it should be checked to ensure chemical compatibility.

New Word Review

Active ingredient—the part of a product that actually does what the product is designed to do. It is not necessarily the largest part of the product. For example, an insecticidal spray may contain less than 1% pyrethrin, the ingredient that actively kills insects. The remaining ingredients are often called inert ingredients.

Acute—sudden or brief, used to describe either an exposure or a health effect. An acute exposure is a short-term exposure. Short-term means lasting for minutes, hours, or days. An acute health effect develops either immediately or a short time after an exposure.

Autoignition temperature—the lowest temperature at which materials begin to burn in air in the absence of a spark or flame. Many chemicals will decompose (break down) when heated. The autoignition temperature is the temperature at which the chemicals formed by decomposition begin to burn. Autoignition temperatures for a specific material can vary by one hundred degrees Celsius or more, depending on the test method used. Therefore, values listed on the MSDS may be rough estimates. To avoid the risk of fire or explosion, materials must be stored and handled at temperatures well below the autoignition temperature.

Boiling point—the temperature at which the material changes from a liquid to a gas. Below the boiling point, the liquid can evaporate to form a vapor. As the material approaches the boiling point, the change from liquid to vapor is rapid and vapor concentration in the air can be extremely high. Airborne gases and vapors may pose fire, explosion, and health hazards. Sometimes, the boiling point is given as a range of temperatures, because different ingredients in a mixture can boil at different temperatures. This is the case because different ingredients in a mixture can boil at different temperatures. If the material decomposes (breaks downs) without boiling, the temperature at which it decomposes may be given with the abbreviation *dec.*

Carcinogen—a substance that can cause cancer. *Carcinogenic* means able to cause cancer. *Carcinogenicity* is the ability of a substance to cause cancer. Under the Controlled Products regulations, materials are identified as carcinogenic if they are recognized as carcinogens by the American Conference of Governmental Industrial Hygienist (ACGIH), or the International Agency for Research on Cancer (IARC). MSDSs from the United States also identify carcinogens recognized by the U.S. National Toxicology Program (NTP). The lists of carcinogens prepared by these organizations include known human carcinogens and some materials that cause cancer in animal experiments. Certain chemicals may be listed as suspect or possible carcinogens if the evidence is limited or so variable that a definite conclusion cannot be made.

CAS Registry Number—a number assigned to a material by the Chemical Abstract Service (CAS) to provide a single unique identifier. A unique identifier is necessary because the material can have many different names. For example, the name given to a specific chemical may vary from one language or country to another. The CAS Registry Number has no significance in terms of the chemical nature or hazards of the material.

Ceiling (C)—the concentration that should not exceeded at any time (the exposure limit).

Chemical name—the proper scientific name for the principal or active ingredient of the product. For example, the chemical name for the herbicide 2.4-D is 2.4-dichlorophenoxyacetic acid. The chemical name can be used to obtain additional information.

Chronic—long-term or prolonged, describing either an exposure or a health effect. A chronic exposure is a long-term exposure. Long-term means lasting for months or years. A chronic health effect is an effect that appears months or years after an exposure. The Controlled Products Regulations describe technical criteria for identifying materials that cause chronic health effects.

Combustible—able to burn. Broadly speaking, a material is combustible if it can catch fire and burn. The terms *combustible* and *flammable* both describe the ability of material to burn. Commonly, combustible materials are less easily ignited than flammable materials.

Compressed gas—under 29 CFR 1910.1200(c), a gas or mixture of gases having, in a container, an absolute pressure exceeding 40 psi at 70°F (21.1°C); a gas or mixture of gases having, in a container, an absolute pressure exceeding 104 psi at 130°F (54.4°C) regardless of the pressure at 70°F; or a liquid having a vapor pressure exceeding 40 psi at 100°F (37.8°C) as determined by ASTM D-323-72. Regardless of whether a compressed gas is packaged in an aerosol can, a pressurized cylinder, or a refrigerated container, it must be stored and handled very carefully. Puncturing or damaging the container or allowing the container to become hot may result in an explosion.

Corrosive material—a material that can attack (corrode) metals or human tissue such as the skin or eyes. Corrosive material can cause metal containers or structural to become weak and eventually to leak or collapse. Corrosive materials can burn or destroy human tissues on contact and can cause effects such as permanent scarring or blindness.

Evaporation rate—a measure of how quickly the material becomes a vapor at normal room temperature. Usually, the evaporation rate is given in comparison to certain chemicals, such as butyl acetate, which evaporates fairly quickly. For example, the rate might be given as "0.5 grams of material evaporates during the same time that 1 gram of butyl acetate evaporates." Often, the evaporation rate is given only as greater or less than 1, which means the material evaporates faster or slower than the comparison chemical. In general, a hazardous material with a higher evaporation rate presents a greater hazard than a similar compound with a lower evaporation rate.

Exposure limit—the concentration of a chemical in the workplace to which most people can be exposed without experiencing harmful effects. Exposure limits should not be taken as sharp dividing lines between safe and unsafe exposures. It is possible for a chemical to cause health effects, in some people, at concentrations lower than the exposure limits. Exposure limits have different names and different meaning, depending on who developed them and whether or not they are legal limits.

Flammable—able to ignite and burn readily. *Flammability* is the ability of a material to ignite and burn readily.

Flash point—the lowest temperature at which a liquid or a solid gives enough vapors to form flammable air-vapor mixture near its surface. The lower the flash point, the greater the fire hazard. The flash point is an approximate value and should not be taken as a sharp dividing line between safe and hazardous conditions. The flash point is determined by a variety of test methods which give different results.

Incompatible materials—materials that can react with the product or with components of the product and may destroy the structure or function of a product; cause a fire, explosion, or violent reactions, or cause the release of hazardous chemicals.

Ingestion—taking a material into the body by mouth (swallowing).

Inhalation—taking a material into the body by breathing it in.

Melting point—the temperature at which a solid material becomes a liquid. The *freezing point* is the temperature at which a liquid material becomes a solid. Usually one value or the other is given on an MSDS. It is important to know the melting or freezing point for storage or handling purposes. For example, a melted or frozen material may burst a container. Also, a change of physical state could alter the hazards of the materials.

Oxidizing material—a material that gives up oxygen easily or can readily oxidize other materials. Examples of oxidizing agents are chlorine and peroxide compounds. These chemicals will support a fire and are highly reactive.

ppm—parts per million. It is a common unit of concentration of gases or vapor in air. For example, 1 ppm of a gas means that 1 unit of gas is present for every 1 million units of air.

Sensitization—the development, over time, of an allergic reaction to a chemical. The chemical may cause a mild response on the first few exposures but, as the allergy develops, the response becomes worse with subsequent exposures. Eventually, even short exposures to low concentration can cause very severe reaction. There are two different types of occupational sensitization: skin and respiratory. Typical symptoms of skin sensitivity are swelling, redness, itching, pain, and blistering. Sensitization of the respiratory system may result in symptoms similar to a severe asthma attack. These symptoms include wheezing, difficulty in breathing, chest tightness, coughing, and shortness of breath.

STEL (Short-term Exposure Limit)—the average concentration to which workers can be exposed to for a short period (usually 115 minutes) without experiencing irritation, long-term or irreversible tissue damage, or reduced alertness. The number of times the concentration reaches the STEL and the amount of time between these occurrences can also be restricted.

Synergism—as used on MSDSs, a term that means that exposure to more than one chemical can result in health effects greater than those expected when the effects of exposure to each chemical are added together. Very simply, it is like saying 1 + 1 = 3. When chemicals are *synergistic*, the potential hazards of the chemicals should be reevaluated, taking their synergistic properties into consideration.

Toxic—able to cause harmful health effects. *Toxicity* is the ability of a substance to cause harmful health effects. Description of toxicity (e.g., low, moderate, severe, etc.) depends on the amount needed to cause an effect or the severity of the effect.

TWA (Time-Weighted Average)—said of exposure limits, the average concentration of a chemical in air for a normal 8-hour workday and 40-hour workweek to which nearly all workers may be exposed day after day without harmful effects. Time-weighted average means that the average concentration has been calculated using the duration of exposure to different concentrations of the chemical during a specific time period. In this way, higher and lower exposures are averaged over the day or week.

Vapor density—the mass per unit volume of a pure gas or vapor. On an MSDS, the vapor density is commonly given as a ratio of the density of the gas or vapor to the density of air. The density of air is given a value of 1. Light gases (density less than 1) such as helium rise in air. If there is inadequate ventilation, heavy gases and vapors (density greater than 1) can accumulate in low-lying areas such as pits and along floors.

Vapor pressure—the pressure of a vapor when in equilibrium with its liquid or solid form. It is a measure of the tendency of a material to form a vapor. The higher the vapor pressure, the higher the potential vapor concentration. In general, a material with a high vapor pressure is more likely to be an inhalation or fire hazard than a similar material with a lower vapor pressure.

Chapter Review Questions

7.1. MSDS sheets may be written in any format, but must contain the information found in OSHA Form _____.

7.2. Section (g)(2)(I)(C) of the 1910.1200 deals with MSDS for _____.

7.3. MSDS should be consulted _____ working with a chemical.

7.4.   A short term exposure: _____.

7.5. _____ means able to burn.

7.6. _____ is the average concentration of a chemical in air for a normal 8-hour workday and 40-hour workweek.

7.7. _____ is the lowest temperature at which a liquid or a solid gives off enough vapors to form flammable air-vapor mixture near its surface.

7.8. Temperature at which a solid material becomes a liquid: _____.

7.9. _____ is unit of gas is present for every 1 million units of air.

7.10. 1 + 1 = 3: _____.

7.11. 12 mg/l of calcium in water expresses roughly the same concentration as _____ calcium in water.

7.12. 14 mg/l of sodium chloride in water expresses roughly the same concentration as _____ salt in water.

Notes

1. OSHA. *Hazard Communication.* www.osha.gov/SLTC/hazardcommunications/index.html (accessed December 15, 2002).

2. Ibid.

3. OSHA has developed a nonmandatory format, OSHA Form 174, which may be used by chemical manufacturers and importers to comply with the rule.

4. 29 CFR 1910.1200(b)(2). While manufacturers and importers must provide the MSDS in English, this requirement does not prohibit a manufacturer, importer, or employer from also providing translated versions of the MSDS to assist non-English-speaking employees in understanding the health and physical hazards to which they are exposed in the workplace.

5. 29 CFR 1910.1200.

CHAPTER 8

Chemistry of Water and Water Pollution

We're all downstream.

—Ecologists' motto

Topics in This Chapter

- Water Chemistry Fundamentals
- Colligative Properties
- Colloids/Emulsions
- Water Constituents
- Important Properties of Water
- The Chemistry of Water Pollution
- TMDLs
- DDT and Biomagnification
- Chemistry and the Clean Water Act
- Wastewater Treatment

In many ways, environmental chemistry revolves around the chemistry of water. Aspects of chemical analysis for which environmental practitioners may be responsible when dealing with water quality samples are:

- Following Standard Operating Procedures (SOP) to conduct chemical assays, including, but not limited to, total Kjeldahl-N (TKN), iron, particulate-P, nitrate, ammonium, dissolved and total P, and total suspended sediments
- Analyzing samples and performing assays on special quality control samples and method detection limit studies
- Developing assay methods to meet specific needs
- Conducting quality control by performing preliminary evaluation of Q tests
- Informing Quality Assurance Officer of abnormal results

This chapter is based in part on F. R. Spellman, *The Science of Water: Concepts and Applications* (Boca Raton, Fla., CRC Press, 1998) and *The Science of Environmental Pollution* (CRC Press, 1990), and F. R. Spellman and N. Whiting, *Water Pollution Control Technology* (Rockville, Md.: Government Institutes, 1999).

- Making data entry and managing sample check-ins, custody forms processing; clarifying bottle labeling and storing samples; prioritizing assay scheduling, designing assay worksheets, and recording/verifying assay results

Because of all these requirements, the environmental professional responsible for water quality compliance must possess skills in chemistry and laboratory techniques, specifically:

- Standards preparation, pipetting, filtration, correct use of volumetric glassware, titration methods, weighing, and following written lab procedures
- Knowledge of theory and methods of low-level nutrient concentrations
- Ability to conduct reagent preparation and purification; knowledge of potential sources of contamination
- Knowledge of lab instrumentation

Aside from these basic skills, the environmental professional actually needs a command of much, much more: a fundamental knowledge of all facets of water chemistry.

Water Chemistry Fundamentals

Whenever we add a chemical substance to another chemical substance, such as adding sugar to tea or adding chlorine to water to make it safe to drink, we are performing the work of chemists. We are working as chemists because we are working with chemical substances, and how they react is important to us.

On the job, we may be required to add chemicals to certain industrial unit processes, for example, in water treatment operations. Table 8.1 lists some of the chemicals and their common applications in water treatment operations.

THE WATER MOLECULE

Just about everyone knows that water is a chemical compound of two simple and abundant elements—hydrogen and oxygen (H_2O). Yet scientists continue to argue the merits of rival theories on the structure of water. The fact is that we still know little about water. For example, *we don't know how water works*.

The reality is that water is very complex. Water has many unique properties that are essential to life and determine its environmental chemical behavior. The water molecule is "different." The two hydrogen atoms (the two in the H_2 part of the water formula) *always* come to rest at an angle of approximately 105° from each other. The hydrogens tend to be positively charged and the oxygen tends to be negatively charged. This arrangement gives the water molecule an electrical polarity; that is, one end is positively charged and one end negatively charged. This 105° relationship makes water lopsided, peculiar, and eccentric—it breaks all the rules (figure 8.1).

Table 8.1.   Chemicals and chemical compounds used in water treatment.

Name	Common Application	Name	Common Application
Activated carbon	Taste and odor control	Aluminum sulfate	Coagulation
Ammonia	Chloramine disinfection	Ammonium sulfate	Coagulation
Calcium hydroxide	Softening	Calcium hypochlorite	Disinfection
Calcium oxide	Softening	Carbon dioxide	Recarbonation
Copper sulfate	Algae control	Ferric chloride	Coagulation
Ferric sulfate	Coagulation	Magnesium hydroxide	Defluoridation
Oxygen	Aeration	Potassium permanganate	Oxidation
Sodium aluminate	Coagulation	Sodium bicarbonate	pH adjustment
Sodium carbonate	Softening	Sodium chloride	Ion exchanger regeneration
Sodium fluoride	Fluoridation	Sodium fluosilicate	Fluoridation
Sodium hexametaphosphate	Corrosion control	Sodium hydroxide	pH adjustment
Sodium hypochlorite	Disinfection	Sodium silicate	Coagulation aid
Sodium thiosulfate	Dechlorination	Sulfur dioxide	Dechlorination
Sulfuric acid	pH adjustment		

In the laboratory, pure water contains no impurities, but in nature water contains a lot of materials besides water. This is an important consideration for the environmental specialist tasked with maintaining the purest, cleanest water possible.

Water is often called the *universal solvent*, and this is fitting, when you consider that, given enough time, water will dissolve anything and everything on earth.

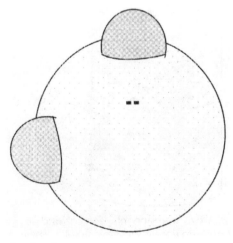

Figure 8.1.   A molecule of water.

WATER SOLUTIONS

A *solution* is a condition in which one or more substances are uniformly and evenly mixed or dissolved. In other words, a solution is a homogeneous mixture of two or more substances. Solutions can be solids, liquids, or gases, such as drinking water, seawater, air, etc. We focus primarily on liquid solutions.

A solution has two components: a *solvent* and a *solute* (see figure 8.2). The solvent is the component that does the dissolving. Typically the solvent is the species present in the greater quantity. The solute is the component that is dissolved. When water dissolves substances, it creates solutions with many impurities.

Generally, a solution is usually transparent and not cloudy and visible to longer-wavelength ultraviolet light. Since water is colorless, the light necessary for photosynthesis can travel to considerable depths. However, a solution may be colored when the solute remains uniformly distributed throughout the solution and does not settle with time.

When molecules dissolve in water, the atoms making up the molecules come apart (dissociate) in the water. This dissociation in water is called *ionization*. When the atoms in the molecules come apart, they do so as charged atoms (both negatively and positively charged), which are called ions. The positively charged ions are called *cations* and the negatively charged ions are called *anions*. Here are some examples of ionization:

$CaCO_3$ (calcium carbonate) $\rightarrow$ Ca^{++} (calcium ion: cation)
$+ CO_3^{-2}$ (carbonate ion: anion)
$NaCl$ (sodium chloride) $\rightarrow$ Na^+ (sodium ion: cation)
$+ Cl^-$ (chloride ion: anion)

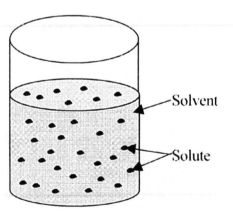

Figure 8.2. Solution with two components: solvent and solute. Source: *Spellman's Standard Handbook for Wastewater Operators*, vol. 1, by Frank Spellman (Technomic, 1999).

Some of the common ions found in water are:

- Hydrogen (H^+)
- Sodium (Na^+)
- Potassium (K^+)
- Chloride (Cl^-)
- Bromide (Br^-)
- Iodide (I^-)
- Bicarbonate (HCO_3^-)

Solutions serve as a vehicle to (1) allow chemical species to come into close proximity so that they can react; (2) provide a uniform matrix for solid materials, such as paints, inks, and other coatings so that they can be applied to surfaces; and (3) dissolve oil and grease so that they can be rinsed away.

Water dissolves polar substances better than nonpolar substances. For example, *polar substances* such as mineral acids, bases, and salts are easily dissolved in water. *Nonpolar substances* such as oils and fats and many *organic* compounds *do not* dissolve as easily in water.

CONCENTRATIONS

Since the properties of a solution depend largely on the relative amounts of solvent and solute, the *concentrations* of each must be specified.

✔ **Key Point:** Chemists use both relative terms, such as saturated and unsaturated, as well as more exact concentration terms, such as weight percentages, molarity, and normality.

Though polar substances dissolve better than nonpolar substances in water, polar substances dissolve in water only to a point; that is, only so much solute will dissolve at a given temperature. When that limit is reached, the resulting solution is *saturated*: no more solute can be dissolved. A liquid/solids solution is *supersaturated* when the solvent actually dissolves more than an equilibrium concentration of solute (usually when heated).

Specifying the relative amounts of solvent and solute, or specifying the amount of one component relative to the whole usually gives the exact concentrations of solutions. Solution concentrations are sometimes specified as *weight percentages*.

$$\% \text{ of solute} = \frac{\text{mass of solute}}{\text{total mass of solution}} \times 100$$

To understand the concepts of *molarity*, *molality*, and *normality*, we must first understand the concept of a *mole*. The mole is defined as the amount of a substance that contains exactly the same number of items (i.e., atoms, molecules, or ions) as 12 grams of carbon-12. By experiment, Avogadro determined this number to be 6.02×10^{23} (to three significant figures). If one mole of C atoms equals 12 g, how much is the mass of one mole of H atoms?

1. Note that carbon is $12\times$ as heavy as hydrogen.
2. Therefore we need only one-twelfth the weight of H to equal the same number of atoms of C (see figure 8.3).

✔ **Key Point:** One mole of H equals one gram.

By the same principle:

- One mole of CO_2 = $12 + 2(16)$ = 44g
- One mole of Cl^- = 35.5g
- One mole of Ra = 226g

In other words, we calculate the mass of a mole if we know formula of the "item."
   Molarity (M) is defined as the number of moles of solute per *liter* of *solution*. The volume of a solution is easier to measure in the lab than its *mass*.

$$M = \frac{\text{no. of moles of solute}}{\text{no. of liters of solution}}$$

Molality (m) is defined as the number of moles of solute per *kilogram* of *solvent*.

$$m = \frac{\text{no. of moles of solute}}{\text{no. of kilograms of solution}}$$

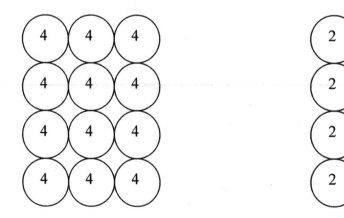

12 items   **12 items**
Mass =48   **Mass = 24**

Figure 8.3. Twice as heavy, so twice the mass in order to equal the same number of items.

✔ **Key Point:** Molality is not as frequently used as molarity, except in theoretical calculations.

Especially for acids and bases, the *normality* (N) rather than the *molarity* of a solution is often reported.

$$N = \frac{\text{no. of equivalents of solute}}{\text{no. of liters of solution}}$$

In acid/base terms, an *equivalent* (or gram equivalent weight) is the amount that will react with one mole of H^+ or OH^-. For example:

1 mole of HCl will generate one mole of H^+, therefore 1 mole HCl = 1 equivalent
1 mole of $Mg(OH)_2$ will generate *two* moles of OH^-, therefore 1 mole of $Mg(OH)_2$
  = 2 equivalents
$HCl \rightarrow H^+ + Cl^-$
$Mg(OH)^{+2} \rightarrow Mg^{+2} + 2OH^-$

By the same principle:

A 1M solution of H_3PO_4 is 3N
A 2N solution of H_2SO_4 is 1M
A 0.5N solution of NaOH is 0.5M
A 2M solution of HNO_3 is 2N

Chemists *titrate* (i.e., determine the concentration of a solution by adding a solution of a reactant to a solution of sample until an indicator changes color) acid/base solutions to determine their normality. An *endpoint indicator* is used to identify the point at which the titrated solution is neutralized.

✔ **Key Point:** If it takes 100 ml of 1N HCl to neutralize 100 ml of NaOH, then the NaOH solution must also be 1N.

PREDICTING SOLUBILITY

Predicting solubility is difficult, but again, there is that rule of thumb: *Like dissolves like.*

Liquid–Liquid Solubility

Liquids with similar structure, and hence similar *intermolecular* forces, will be completely miscible. For example, we would correctly predict that methanol and water are completely soluble in any proportion (see figure 8.4).

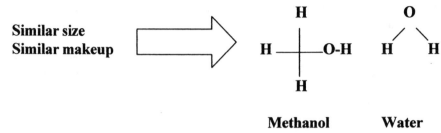

Similar size
Similar makeup

Methanol   Water

Figure 8.4.   Predicting solubility, liquid to liquid.

Liquid–Solid Solubility

Solids *always* have limited solubilities in liquids, in general because of the difference in magnitude of their intermolecular forces. Therefore, the closer the temperature is to its melting point, the better the match between a solid and a liquid.

✔ **Key Point:** At a given temperature, lower melting solids are more soluble than higher melting solids.

Structure is also important; for example, nonpolar solids are more soluble in nonpolar solvents.

Liquid–Gas Solubility

As with solids, the more similar the intermolecular forces, the higher the solubility.

Therefore, the closer the match between the temperature of the solvent and the boiling point of the gas, the higher the solubility.

When water is the solvent, an additional *hydration* factor promotes solubility of charged species (see figure 8.5).

Other factors that can significantly affect solubility are *temperature* and *pressure*. In general, raising the temperature typically increases the solubility of solids in liquids.

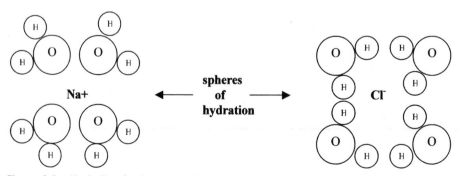

Figure 8.5.   Hydration factor promoting solubility of charged species.

✔ **Key Point:** Dissolving a solid in a liquid is usually an *endothermic process* (i.e., heat is absorbed), so raising the temperature will "fuel" this process.

In contrast, dissolving a gas in a liquid is usually an *exothermic* process (i.e., it evolves heat). Therefore lowering the temperature generally *increases* the solubility of gases in liquids.

✔ **Interesting Point:** "Thermal" pollution is a problem because of the decreased solubility of O_2 in water at higher temperatures.

Pressure has an appreciable effect only on the solubility of *gases* in liquids. For example, carbonated beverages such as soda water are typically bottled at significantly higher atmospheres. When the beverage is opened, the decrease in the pressure above the liquid causes the gas to bubble out of solution. When shaving cream is used, dissolved gas comes out of solution, bringing the liquid with it as foam.

Colligative Properties

The *colligative properties* of a solution are those properties that depend only on the concentration of a solute species. The point is that some properties of a solution depend on the *concentrations* of the solute species rather than their identity. These include:

• Lowering vapor pressure
• Raising boiling point
• Decreasing freezing point
• Osmotic pressure

True colligative properties are directly *proportional to the concentration* of the solute but entirely independent of its identity.

LOWERING OF VAPOR PRESSURE

With all other conditions identical, the vapor pressure of water above the pure liquid is higher than that above sugar water. The vapor pressure above a 0.2 m sugar solution is the same as that above a 0.2 m urea solution. The lowering of vapor pressure above a 0.4 m sugar solution is twice as great as that above a 0.2 m sugar solution.

Solutes lower vapor pressure because they *lower the concentration of solvent* molecules. To remain in equilibrium, the solvent vapor concentration must decrease (hence the vapor pressure decreases).

RAISING THE BOILING POINT

A solution containing a *nonvolatile* solute boils at a higher temperature than the pure solvent. The increase in boiling point is directly proportional to the increase in solute concentration in dilute solutions. This phenomenon is explained by the lowering of vapor pressure already described.

DECREASING THE FREEZING POINT

At low solute concentrations, solutions generally freeze or melt at temperatures lower than the freezing point of the pure solvent.

✔ **Key Point:** The presence of dissolved "foreign bodies" tends to interfere with freezing, and therefore solutions can only be frozen at temperatures below that of the pure solvent.

✔ **Key Point:** We add antifreeze to the water in our radiators to both lower its freezing point and increase its boiling point.

OSMOTIC PRESSURE

Water moves spontaneously from an area of high vapor pressure to an area of low vapor pressure (see figure 8.6).

   If this experiment were allowed to continue, in the end all of the water would move to the solution (see figure 8.7).

   A similar process will occur when pure water is separated from a concentrated solution by a *semipermeable* membrane (i.e., it only allows the passage of water molecules; see figure 8.8).

   The *osmotic pressure* is the pressure that is just adequate to prevent osmosis. In

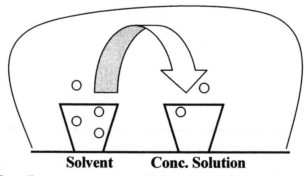

Solvent   Conc. Solution

Figure 8.6. Osmotic pressure, movement into concentrated solution.

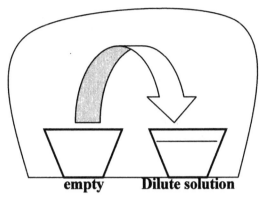

Figure 8.7. Osmotic pressure, movement into dilute solution.

dilute solutions, the osmotic pressure is directly proportional to the solute concentration and is independent of its identity (see figure 8.9).

The properties of electrolyte solutions follow the same trends as nonelectrolyte solutions, but are also dependent on both the *nature* of the electrolyte as well as its *concentration.*

$NaCl$
Na_2SO_4
$CaCl_2$
$MgSO_4$

Colloids/Emulsions

A *solution* is a homogeneous mixture of two or more substances (e.g., seawater). A *suspension* is a brief commingling of solvent and undissolved particles (e.g., sand and water). A *colloidal suspension* is a commingling of particles not visible to the naked eye but larger than individual molecules.

✔ **Key Point:** Colloidal particles do not settle out by gravity alone.

Colloidal suspensions can consist of:

- *Hydrophilic* "solutions" of macromolecules, such as proteins, that spontaneously form in water
- *Hydrophobic* suspensions, which gain stability from their repulsive electrical charges (see figure 8.10)
- *Micelles,* which are special colloids having charged hydrophilic "heads" and long hydrophobic "tails" (see figure 8.11)

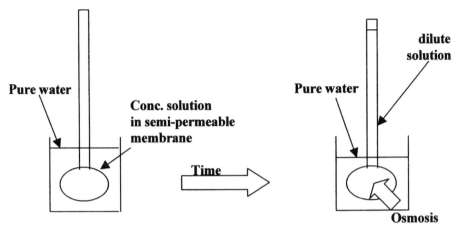

Figure 8.8.   Colligative properties: passage of water molecules only.

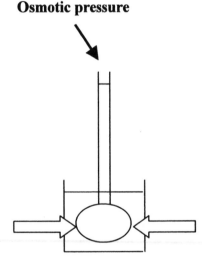

Figure 8.9.   Osmotic pressure: pressure adequate enough to prevent osmosis.

Colloids are usually classified according to the original states of their constituent parts (see table 8.2). The *stability* of colloids can be primarily attributed to *hydration* and *surface charge,* both of which help to prevent contact and subsequent coagulation.

✔ **Key Point:** In many cases, water-based *emulsions* have been used to replace organic solvents (e.g., paints, inks, etc.), even though the compounds are not readily soluble in water.

In wastewater treatment, the *elimination* of colloidal species and emulsions is achieved by various means, including:

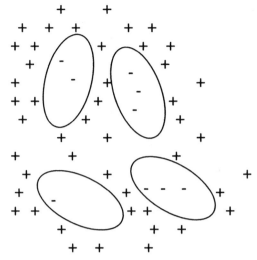

Figure 8.10.   Hydrophobic colloidal suspension.

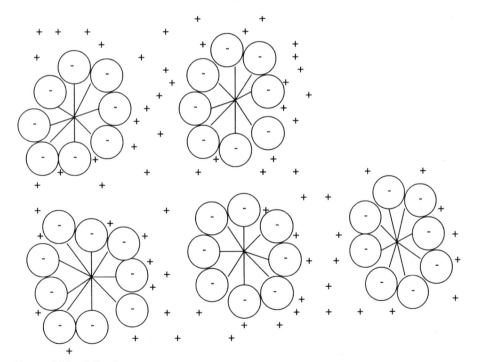

Figure 8.11.   Micelles.

Table 8.2.   Types of colloids.

Name	Dispersing Medium	Dispersed Phase
Solid sol	Solid	Solid
Gel	Solid	Liquid
Solid form	Solid	Gas
Sol	Liquid	Solid
Emulsion	Liquid	Liquid
Foam	Liquid	Gas
Solid aerosol	Gas	Solid
Aerosol	Liquid	Aerosol

Source: Types of Colloids, www.ch.bris.ac.uk/webprojects2002/pdavies.types.html (accessed December 18, 2002).

- Agitation
- Heat
- Acidification
- Coagulation (adding ions)
- Flocculation (adding bridging groups)

Water Constituents

Natural water can contain a number of substances, or what we call impurities or *constituents*. When a particular constituent can affect the good health of the water user, it is called a *contaminant* or *pollutant*. It is these contaminants that the environmental specialist works to prevent entering or to remove from the water supply.

SOLIDS

Other than gases, all contaminants of water contribute to the *solids* content. Natural waters carry a lot of dissolved solids and *not-dissolved solids*—nonpolar substances and relatively large particles of materials, such as silt, that won't dissolve. Classified by their size and state, by their chemical characteristics, and their size distribution, solids can be dispersed in water in both suspended and dissolved forms. In regards to size, solids in water can be classified as:

- Suspended solids
- Settleable solids
- Colloidal solids
- Dissolved solids

Total solids are those solids, both suspended and dissolved, that remain behind when the water is removed by evaporation. Solids are also characterized as being *volatile* or *nonvolatile*.

✔ **Important Point:** Though not technically accurate from a chemical point of view because some finely suspended material can actually pass through the filter, suspended solids are defined as those that can be filtered out in the suspended-solids laboratory test. The material that passes through the filter is defined as *dissolved solids*.

Colloidal solids are extremely fine suspended solids (particles) of less than one micron in diameter; they are so small they will not settle even if allowed to sit quietly for days or weeks (but they still make water cloudy).

TURBIDITY

One of the first characteristics people notice about water is its clarity. *Turbidity* is a condition in water caused by the presence of suspended matter, resulting in the scattering and absorption of light rays. In plain English, turbidity is a measure of the light-transmitting properties of water. Natural water that is very clear (low turbidity) allows one to see images at considerable depths. High-turbidity water appears cloudy. Even water with low turbidity, however, can still contain dissolved solids. Dissolved solids do not cause light to be scattered or absorbed; thus, the water looks clear. High turbidity causes problems for the waterworks operator because the components that cause high turbidity can cause taste and odor problems and will reduce the effectiveness of disinfection.

COLOR

Color is considered an aesthetic quality of water, with no direct health impact. However, often the color of water can be deceiving. Many of the colors associated with water are not "true" colors but the result of colloidal suspension (apparent color). This apparent color can be attributed to dissolved tannin extracted from decaying plant material. *True color* is the result of dissolved chemicals, most often organics that cannot be seen.

DISSOLVED OXYGEN (DO)

Gases, notably oxygen, carbon dioxide, hydrogen sulfide, and nitrogen, can be dissolved in water. Gases dissolved in water are significant. Carbon dioxide, for example, is important because of the role it plays in pH and alkalinity. Carbon dioxide is released into the water by microorganisms and consumed by aquatic plants. Dissolved oxygen (DO) in water is of most importance to waterworks operators because it is an indicator of water quality.

We stated earlier that solutions could become saturated with solute. This is also the case with water and oxygen. The amount of oxygen that can be dissolved at saturation depends upon the temperature of the water. However, in the case of oxygen, the effect is just the opposite of other solutes. The higher the temperature the lower the saturation level; the lower the temperature the higher the saturation level.

METALS

One of the constituents or impurities often carried by water is *metal*. At normal levels, most metals are not harmful, but a few metals can cause taste and odor problems in drinking water. Some metals may be toxic to humans, animals, and microorganisms. Most metals enter water as part of compounds that ionize to release the metal as positive ions. Table 8.3 lists some metals commonly found in water and their potential health hazards.

ORGANIC MATTER

Organic matter or *organic compounds* are those that contain the element *carbon* and are derived from material that was once alive (i.e., plants and animals), such as:

• Fats
• Dyes
• Soaps
• Rubber products
• Wood
• Fuels
• Cotton
• Proteins
• Carbohydrates

Organic compounds in water are usually large, nonpolar molecules that do not dissolve well in water. They often provide large amounts of energy to animals and microorganisms.

Table 8.3. Common metals found in water.

Metal	Health Hazard
Barium	Circulatory system effects and increased blood pressure
Cadmium	Concentration in the liver, kidneys, pancreas, and thyroid
Copper	Nervous system damage and kidney effects, toxic to humans
Lead	Same as copper
Mercury	Central nervous system disorders
Nickel	Central nervous system disorders
Selenium	Central nervous system disorders
Silver	Turns skin gray
Zinc	Causes taste problems but not a health hazard

INORGANIC MATTER

Inorganic matter or *inorganic compounds* are carbon-free, not derived from living matter, and easily dissolved in water; they are of mineral origin. The inorganics include acids, bases, oxides, salts, etc. Several inorganic components are important in establishing and controlling water quality.

ACIDS

An *acid* is a substance that produces hydrogen ions (H^+) when dissolved in water. Hydrogen ions are hydrogen atoms stripped of their electrons. A single hydrogen ion is nothing more than the nucleus of a hydrogen atom. Lemon juice, vinegar, and sour milk are acidic (contain acid). The common acids used in treating water are hydrochloric acid (HCl), sulfuric acid (H_2SO_4), nitric acid (HNO_3), and carbonic acid (H_2CO_3). Note that in each of these acids, hydrogen (H) is one of the elements. The relative strengths of acids in water, listed in descending order, are classified in table 8.4.

BASES

A *base* is a substance that produces hydroxide ions (OH^-) when dissolved in water. Lye or common soap (bitter things) contains bases. Bases used in waterworks operations are calcium hydroxide ($Ca(OH)_2$), sodium hydroxide ($NaOH$), and potassium hydroxide (KOH). Note that the hydroxyl group (OH) is found in all bases. Certain bases also contain metallic substances, such as sodium (Na), calcium (Ca), magnesium (Mg), and potassium (K). These bases contain the elements that produce the alkalinity in water.

Table 8.4.   Relative strengths of acids in water.

Acid	Chemical Symbol
Perchloric acid	$HClO_4$
Sulfuric acid	H_2SO_2
Hydrochloric acid	HCl
Nitric acid	HNO_3
Phosphoric acid	H_3PO_4
Nitrous acid	HNO_2
Hydrofluoric acid	HF
Acid	CH_3COOH
Carbonic acid	H_2CO_3
Hydrocyanic acid	HCN
Boric acid	H_3BO_3

SALTS

When acids and bases chemically interact, they neutralize each other. The compound other than water that forms from the neutralization of acids and bases is called a *salt*. Salts constitute, by far, the largest groups of inorganic compounds. Copper sulfate, a common salt, is used in waterworks operations to kill algae in water.

pH

pH is a measure of the hydrogen ion (H^+) concentration. Solutions range from very acidic (having a high concentration of H^+ ions) to very basic (having a high concentration of OH^- ions). The pH scale ranges from 0 to 14, with 7 being the neutral value.

The pH of water is important to the chemical reactions that take place within water, and pH values that are too high or low can inhibit growth of microorganisms. High pH values are considered basic and low pH values are considered acidic. Stated another way, low pH values indicate a high level of H^+ concentration, while high pH values indicate a low H^+ concentration. Because of this inverse *logarithmic* relationship, there is a tenfold difference in H^+ concentration.

Natural water varies in pH, depending on its source. Pure water has a neutral pH, with an equal number of H^+ and OH^-. Adding an acid to water causes additional positive ions to be released so that the H^+ ion concentration goes up and the pH value goes down.

$$HCl \rightarrow H^+ + Cl^-$$

Changing the hydrogen ion activity in solution can shift the chemical equilibrium of water. Thus pH adjustment is used to optimize coagulation, softening, and disinfection reactions, and for corrosion control. To control water coagulation and corrosion, it is necessary for the waterworks operator to test for hydrogen ion concentration of the water to get pH. In coagulation tests, as more alum (acid) is added, the pH value is lowered. If more lime (alkali, a base) is added, the pH value is raised. This relationship is important—and if good floc is formed, the pH should then be determined and maintained at that pH value until there is a change in the new water.

ALKALINITY

Alkalinity is defined as the capacity of water to accept protons (positively charged particles); it can also be defined as a measure of water's ability to neutralize an acid. Stated in even simpler terms: Alkalinity is a measure of water's capacity to absorb hydrogen ions (i.e., to neutralize acids) without significant pH change.

Bicarbonates, carbonates, and hydrogen cause alkaline compounds in a raw or treated water supply. Bicarbonates are the major components, because of carbon diox-

Table 8.5.   Classifications used in water hardness.

Classification	mg/l CaCO$_3$
Soft	0–75
Moderately Hard	75–150
Hard	150–300
Very Hard	Over 300

ide action on "basic" materials of soil; borates, silicates, and phosphates may be minor components. Alkalinity of raw water may also contain salts formed from organic acids, such as humic acid. Alkalinity in water acts as a buffer that tends to stabilize and prevent fluctuations in pH. It is usually beneficial to have significant alkalinity in water because it would tend to prevent quick changes in pH. Quick changes in pH interfere with the effectiveness of the common water treatment processes. Low alkalinity also contributes to corrosive tendencies of water. When alkalinity is below 80 mg/l, it is considered low.

HARDNESS

Hardness may be considered a physical or chemical parameter of water. It represents the total concentration of calcium and magnesium ions, reported as calcium carbonate. Hardness causes soaps and detergents to be less effective and contributes to scale formation in pipes and boilers. Hardness is not considered a health hazard. However, lime precipitation or ion exchange must often soften water that contains hardness. Low hardness contributes to the corrosive tendencies of water. Hardness and alkalinity often occur together because some compounds can contribute both alkalinity and hardness ions. Hardness is generally classified as shown in table 8.5.

Important Properties of Water

SOLUBILITY

Compounds that can form hydrogen bonds with water *tend to be far more soluble* in water than compounds that cannot form H-bonds (see figure 8.12).

SURFACE TENSION

Water has a high surface tension. Surface tension governs surface phenomena and is an important factor in physiology.

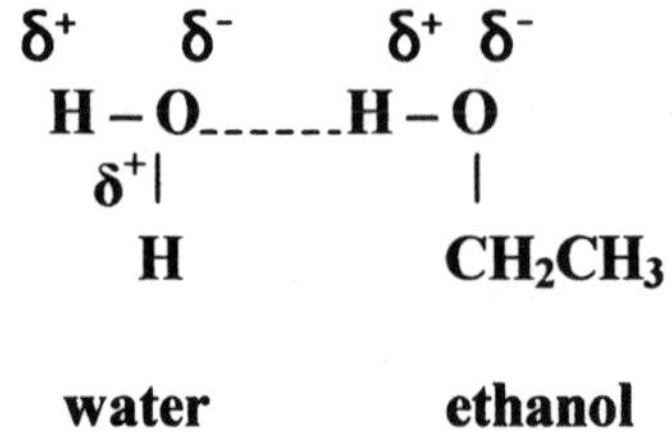

$$\begin{array}{cccc} \delta^+ & \delta^- & \delta^+ & \delta^- \\ H-O & \text{......} & H-O & \\ \delta^+| & & | & \\ H & & CH_2CH_3 & \end{array}$$

| Like |
| Dissolves |
| Like! |

water   ethanol

Figure 8.12.   Compounds soluble in water.

DENSITY

Water has its maximum liquid density at 4°C. Recall that density is mass per unit volume. When water freezes, ice floats.

BOILING POINT

In general, boiling point increases with molecular weight, but *hydrogen bonding* increases the boiling point of water above that predicted based on molecular weight alone.

Figure 8.13 shows water's high "effective" molecular weight.

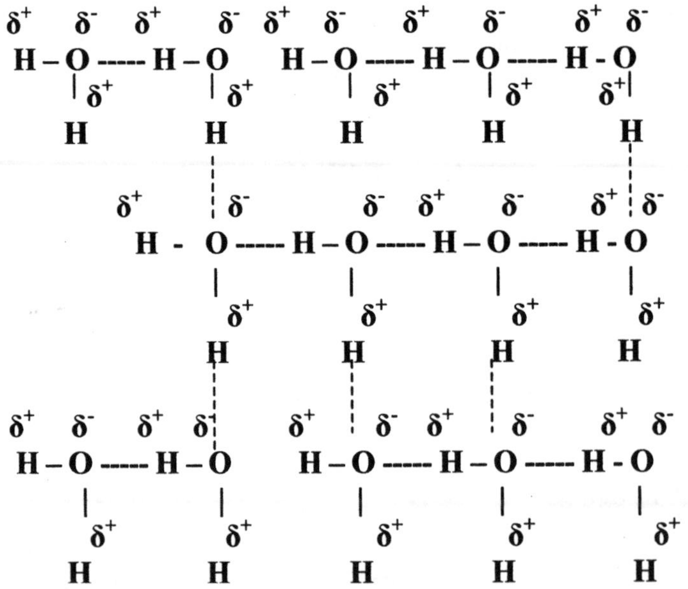

Figure 8.13.   Water's high "effective" molecular weight.

HEAT CAPACITY

Water has a higher heat capacity than any other liquid other than ammonia. Heat capacity is the amount of energy it takes to raise the temperature of a substance one degree. This allows organisms and geographical regions to stabilize temperature more easily.

HEAT OF VAPORIZATION

Water has a higher heat of vaporization than any other material. Heat of vaporization is the energy required to change a liquid to a vapor; it effects the transfer of water molecules between surface water and the atmosphere.

LATENT HEAT OF FUSION

Water has a higher latent heat of fusion than any other liquid other than ammonia. The heat of fusion is the energy released when a liquid condenses to a solid. Temperature is thus stabilized at the freezing point.

PHASE TRANSITIONS OF WATER

Figure 8.14 shows the phase transitions of water. A phase transition is the spontaneous conversion of one phase to another that occurs at a characteristic temperature for a given pressure. For example, at 1 atm, ice is the stable phase of water below 0 degrees C, but above this temperature the liquid is more stable. A phase diagram of water, for example, is a map of the ranges of pressure and temperature at which each phase of the water is the most stable.

The Chemistry of Water Pollution

Providing adequate water supplies of pure drinking water for the world's ever-increasing human population is perhaps the greatest environmental challenge of the twenty-first century. In 1993, more than a hundred residents of Milwaukee were killed and another 400,000 sickened when Cryptosporidium contaminated the city's water supply.

WHAT IS WATER POLLUTION?

"People's opinions differ in what they consider to be a pollutant on the basis of their assessment of benefits and risk to their health and economic well-being. For example,

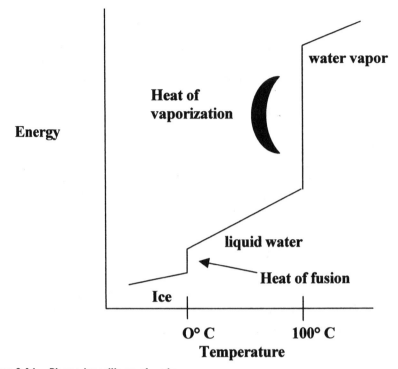

Figure 8.14.   Phase transitions of water.

visible and invisible chemicals spewed into water by an industrial facility might be harmful to people and other forms of life living nearby and in the water body itself. However, if the facility is required to install expensive pollution control equipment, forcing the industrial facility to shut down or to move away, workers who would lose their jobs and merchants who would lose their livelihoods might feel that the risks from polluted air and water are minor weighed against the benefits of profitable employment. The same level of pollution can also affect two people quite differently. Some forms of water pollution, for example, might cause only a slight irritation to a healthy person but cause life-threatening problems to someone with autoimmune deficiency problems. Differing priorities lead to differing perceptions of pollution (concern about the level of pesticides in foodstuffs prompting the need for wholesale banning of insecticides is unlikely to help the starving). Public perception lags behind reality because the reality is sometimes unbearable. Pollution is a judgment, and pollution demands continuous judgment."[1]

ACIDITY, ALKALINITY, AND SALINITY

The acidity of natural water systems is defined as its capacity to neutralize OH^-. Acidity is more difficult to measure than alkalinity because volatile gases such as CO_2 and H_2S affect acidity.

✔ **Key Point:** Common sources of "natural" acidity include weak acids, such as $H_2PO_4^-$, CO_2, H_2S, proteins, fatty acids, and acidic metal ions.

Recall that *alkalinity* is defined as the capacity of water to accept H^+. In contrast to pH, which is an *intensity factor,* alkalinity is a *capacity factor.*

1 liter of 0.001M NaOH has a pH of 11 but can neutralize only 0.001 mole of acid. 1 liter of 0.1M HCO_3^- has a pH of 8.34 but can neutralize 0.1 mole of acid.

Chemists usually express alkalinity in terms or *equivalents per liter.*

$$CaCO_3 + 2H^+ \rightarrow Ca^{+2} + CO_2 + H_2O$$

1 mole of $CaCO_3$ neutralizes 2 moles of H^-
or
0.5 mole of $CaCO_3$ neutralizes 1 mole of H^+
therefore
A 0.5M solution of $CaCO_3$ equals 1 equivalent/L alkalinity

The *salinity* of water is its salt load (i.e., concentration of dissolved salts). Increased salt concentrations can arise from numerous human activities, including:

• Municipal water treatment
• Leaching from waste piles
• Irrigation and agriculture

WATER HARDNESS

Water hardness is attributed to the concentration of Ca^{+2} (plus Mg^{+2} and sometimes Fe^{+2}). As noted above, the degree of water hardness relates to the amount of dissolved minerals, especially calcium and magnesium, in the water. Water hardness is generally expressed in the amount of calcium carbonate ($CaCO_3$). Water hardness is measured in ppm, kH (carbonate hardness), and dH (degrees of hardness) or gH (general hardness). Water is expressed as "soft" (having few dissolved minerals) or "hard" (having many dissolved minerals). General levels of water hardness are expressed in table 8.6 (1 dH is equivalent to about 17 ppm).

High concentrations of dissolved CO_2 enhance the solubility of Ca^{+2}. Heating the water, which drives out CO_2, can decrease the solubility of $Ca+2$. This causes problems in hot water systems, which can become choked or clogged with insolubles.

$$Ca^{+2} + 2HCO_3^- \rightarrow CaCO_3(s) + H_2O$$

In the presence of soap, hard water forms a "curdlike" precipitate.

$$2C_{17}H_{35}COO^-Na^+ + Ca^{+2} \rightarrow Ca(C_{17}H_{35}CO_2)_2(s) + 2Na^+$$

Table 8.6. Water hardness specifications.

(dH)	ppm	(gH)
very soft	0 to 70	0 to 4
soft	70 to 135	4 to 8
medium hard	135 to 200	8 to 12
hard	200 to 350	12 to 20
very hard	over 350	over 20

Although Ca^{+2} does not form insoluble precipitates with detergents, it does adversely affect their performance.

Industrial-scale water-softening techniques (i.e., the removal of Ca^{+2} and Mg^{+2}) include:

1. The addition of lime $Ca(OH)_2$ where only "bicarbonate hardness" is of significant concern:

$$Ca^{+2} + 2HCO_3^- + Ca(OH)_2 \rightarrow 2CaCO_3(s) + 2H_2O$$

2. The addition of lime and soda ash (Na_2CO_3) when bicarbonate is not a factor:

$$Ca^{+2} + 2Cl^- + 2Na^+ CO_3^- \rightarrow CaCO_3(s) + 2Cl^- + 2Na^+$$

3. The conversion of precipitated $CaCO_3$ to lime $Ca(OH)_2$ with heat:

$$CaCO_3(heat) \rightarrow CaO + CO_2(g)$$
$$CaO + H_2O \rightarrow CA(OH)_2$$

Various problems with water "softened" by this last process include residual concentrations of $CaCO_3$ and $Mg(OH)_2$ and extremely high pH.

✔ **Key Point:** Recarbonation generally lowers the pH to an acceptable range.

Additional water softening techniques are ion exchange (see figure 8.15), the addition of orthophosphate (precipitation)

$$5Ca^{+2} + 3PO_4^{-3} + OH^- \rightarrow Ca_5OH(PO_4)_3(s)$$

and chelation (sequestration)

$$Ca^{+2} + Y^{-4} \rightarrow CaY^{-2}$$

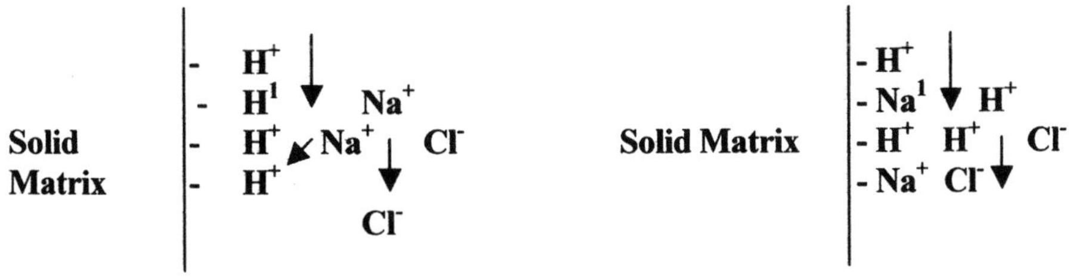

Solid
Matrix

- H^+ ↓
- H^1 ↓ Na^+
- H^+ ↙ Na^+ ↓ Cl^-
- H^+ ↓
        Cl^-

Solid Matrix

- H^+ ↓
- Na^1 ↓ H^+
- H^+ H^+ ↓ Cl^-
- Na^+ Cl^- ↓

Cation exchanger

Figure 8.15.   Ion exchange.

METAL CONTAMINATION AND CHELATING AGENTS

Common aqueous metal contaminants include:

- Common metals (Fe, Mn)
- Heavy metals (Cd, Pb, Hg)
- Metalloids (As, Se, Sb)
- Organometallics
- Radionuclides

Metal ions in water occur as hydrated ions $M(H_2O)_x^{n+}$ or *complexes* (see figure 8.16).

Heavy metals like lead and cadmium are released into the atmosphere by vehicle and power station emissions. These dissolve with water vapor and sulfuric and nitric acids in the atmosphere, returning to the earth as acid rain to pollute water supplies. This problem is compounded by the corrosive effect acids have on metal water pipes, further contaminating water supplies.

Common metal contaminants such as iron and manganese are removed by oxida-

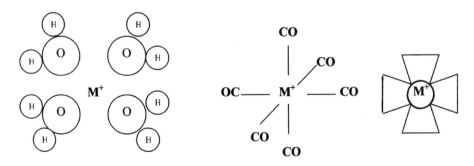

Hydrated metal cation   **Metal complexes**

Figure 8.16.   Complexes: Metal contaminants and chelating agents.

tion to insoluble forms (e.g., Fe(III) and MnO_2). *Colloidal suspensions* or Fe(III) and Mn(IV) and naturally occurring organic substances are difficult to remove.

Water contamination by *heavy metals* (found in the lower right-hand portion of the periodic table) can be very harmful because these elements

- attack S bonds, carboxylic acids, and amino groups;
- bind to cell membranes; and
- precipitate and/or destroy biologically important phosphates.

Cadmium pollution, arising from industrial and mining activities, is thought to replace biological zinc, resulting in serious adverse health effects such as high blood pressure, damage to the kidneys, and destruction of red blood cells.

Lead from food and drinking water contamination, as well as industrial and natural sources, can lead to kidney, brain, and central nervous system disorders, among other serious adverse health effects.

Mercury enters the environment from numerous minor sources, which cumulatively add up to substantial toxic effects:

- Sewage can contain $10 \times$ higher concentration of mercury than natural waters.
- The high levels of mercury found in fish in the early 1970s were traced to the production of soluble CH_3Hg^+ and volatile $(CH_3)_2Hg$, which biomagnified in fish tissue $1000 \times$.

Various methods are used to remove heavy metals. For example, lime treatment removes heavy metals. Other heavy metal removal techniques include:

- Electrodeposition
- Reverse osmosis
- Ion exchange
- Cementation: $Cu^{+2} + Fe$ (iron scrap) $\rightarrow Fe^{+2} + Cu$
- Sorption on activated carbon (+ chelation)
- Ferric chloride flocculation (Cl^3Fe)
- Alum ($Al_2(SO_4)_3^- 18H_2O$) coagulation: $Al(H_2O)_6^{+3} + 3OH^- \rightarrow Al(OH)_3(s) + 6H_2O$

✔ **Important Point:** Caution must be used in the disposal of sludges because of the accumulation of heavy metals and other contaminants removed by precipitation, etc.

Arsenic, which is the primary aqueous metalloid contaminant of concern, is released into the atmosphere in significant amounts from the burning of fossil fuels. Arsenic is also a by-product of lead, copper, and gold refining. Bacteria have also been implicated in the transformation of arsenic into more mobile, toxic forms, such as $(CH_3)_2AsH$.

Organometallics, such as the large class of organotin biocides in current use, are potentially dangerous water pollutants because of their toxicity and widespread use.

Approximately 40,000 metric tons/year of organotin compounds are produced as bactericides, fungicides (anti-fouling), insecticides, and preservatives.

Radionuclides can enter water systems through human and natural sources. Naturally occurring: C-14, Si-32, L-40, Ra-226, Pb-210, Th-230, and Th-234. Significant Ra contamination has been detected in Western mining areas, IA, IL, WI, MS, MN, FL, NC, VA, and New England.

Dangerous radionuclides resulting from human activities include: Sr-90, I-131, Cs-137, Ba-140, Zr-95, Ce-141, Sr-89, Ru-103, Kr-85, Co-60, Mn-54, Fe-55, and Pu-239.

CHELATING AGENTS

If the species complexing with the metal has only one bonding site, it is called a *ligand.* If the species complexing with the metal has multiple bonding sites, it is called a *chelating agent.*

Metals complex with both natural and manmade chelating agents. Natural chelating agents are called *humic substances,* which are the residuals from the biodegradation of vegetation. The molecular weights of these substances range from a few 100 to >10,000.

Soluble humic substances also add a yellow tint to water, solubilize biologically important metals, and generate trihalomethanes (carcinogens) during municipal chlorination processes.

Insoluble humic substances exchange cations with water and accumulate metals.

Manmade chelating agents, such as EDTA, are used as cleaning agents (decontamination) that solubilize metals. This increased metal solubility also increases the mobility of chelated species released into the environment.

Antibiotics such as streptomycin, aspergillic acid, tetracycline and others are known to have chelating properties.

OTHER INORGANIC WATER POLLUTANTS

- NH_3 (from nitrogenous organic wastes)
- H_2S (anaerobic digestion, geothermal emissions, industrial wastes)
- CO_2 (organic decay, recarbonation)
- NO_2^- (corrosion inhibitor)
- SO_3^{-2} (boiler feedwater)
- Asbestos (industrial mining wastes)

The removal of *dissolved inorganics* is an essential part of wastewater treatment, even if the treated water is not destined for drinking water. Distillation is too costly and does not remove inorganic volatiles (e.g., NH_3). Freezing is also not cost effective. Common techniques for the removal of dissolved inorganics include:

- Electrodialysis and reverse osmosis
- Ion exchange
- Air stripping at high pH (NH_3)
- Precipitation (phosphate)

$$5Ca(OH)_2 + 3HPO_4^{-2} \rightarrow Ca_5OH(PO_4)_3(s) + 3H_2O + 6OH^-$$

ORGANIC WATER POLLUTANTS

Domestic and industrial sewage contains a wide variety of organic pollutants, including viruses, detergents, phosphates, grease, oil, salts, heavy metals, chelating agents, solids, and *biorefractory* (i.e., biodegradation resistant) organics.

✔ **Key Point:** If not properly treated, the resulting treated water, as well as sludge, can still contain all or many of these pollutants.

Soaps, *detergents*, *surfactants*, and *detergent builders* are also released into domestic and industrial wastewaters in large quantities (see figure 8.17).

Many manmade organic compounds cannot be easily biodegraded by microorganisms. These low-molecular-weight, relatively nonvolatile biorefractory compounds persist in the environment with unknown consequences. The list of refractory compounds is filled with aromatic and chlorinated hydrocarbons or both.

Biological treatment of biorefractory compounds must be accompanied by other processes including air stripping, solvent extraction, ozonation, and carbon absorption to remove them. Many of these compounds also cause taste and odor problems in drinking water.

Pesticides and *herbicides* are manufactured and used in large quantities and often

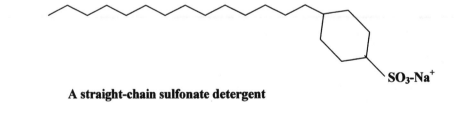

SO_3-Na^+

A straight-chain sulfonate detergent

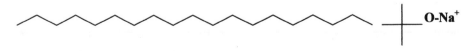

O-Na^+

A straight-chain alkylsulfate detergent

Figure 8.17. Organic water pollutants.

find their way unintentionally into environmental mediums—for example, Kepone contaminated the James River in Virginia. In addition, biorefractory-starting materials for pesticide production, such as hexachlorobenzene, have been found in drinking water. Another example is dioxin, which is a potentially toxic by-product of pesticide and herbicide manufacture.

PCBs also find their way into water bodies. PCBs contain between 1 and 10 CL substitutions of biphenyl, making a total of 209 possible *congeners*. PCBs have been detected in sediments and in animal and bird tissues throughout the world. EPA has estimated that 91% of all Americans have detectable levels of PCBs in their tissues.

Removal of dissolved organics before chlorination is necessary to avoid the formation of trihalomethanes. Organics that typically survive through (or are produced by) secondary waste treatment processes include humic substances ($\sim$ 59%), carbohydrates, detergents, and tannins.

The primary method for the removal of dissolved organics is sorption by activated carbon. The "activation process" creates greater porosity, surface area, and affinity for organics. Activated carbon must be periodically flushed to remove particulates and can be regenerated by heating in steam/air to 950°C.

Other techniques for the removal of dissolved organics include adsorption by synthetic polymers (Amberlite resins) and oxidation.

The *disposal of sludge* from wastewater treatment plants is a major problem around the world. Ocean dumping was completely banned in the United States in 1992. Before acceptable disposal, these sludges are typically reduced in volume by anaerobic digestion, followed by conditioning, thickening, and dewatering.

These nutrient-rich sludges (5% N, 3% P, 0.5% K) can be used to fertilize and condition soil. Problems include contaminated runoff, high concentrations of heavy metals and chemical "precipitating" agents, crop contamination, and pathogens. Sludge samples have contained 9,000 ppm Zn, 6,000 ppm Cu, 600 ppm Ni, and 800 ppm Cd.

TMDLs

TMDLs[2] (Total Maximum Daily Loads) are the amount of a pollutant that can be discharged to a water body and still attain water quality goals. TMDLs are required by law for impaired or polluted waters and are used to set priorities for developing watershed plans and to calculate individual load allocations. Load allocations assign responsibility for water quality to dischargers into the water body.

On July 11, 2000, the Environmental Protection Agency (USEPA) Administrator signed a rule that revises the TMDL program and makes related changes to the National Pollutant Discharge Elimination System (NPDES) and Water Quality Standards programs (65 FR 43585, July 13). According to President Clinton, EPA's move was a "critical, commonsense step" to clean up the nation's waterways.

Why a new TMDL rule? USEPA points out that over 20,000 water bodies across America have been identified as polluted by states, territories, and authorized tribes. These waters include over 300,000 stream/river and shoreline miles and 5 million acres of lakes. The overwhelming majority of the U.S. population lives within 10 miles of these polluted waters.

The Clean Water Act (CWA) provides special authority for restoring polluted waters. The act calls on states to work with interested parties to develop TMDLs for polluted waters. A TMDL is essentially a "pollution budget" designed to restore the health of the polluted body of water.

GOALS OF TMDL RULE

The TMDL rule will make thousands more streams/rivers, lakes, and coastal waters safe for swimming, fishing, and healthy population of fish and shellfish. Key provisions of the TMDL rule include:

1. Requiring states to develop more detailed listing methods and comprehensive lists of polluted water bodies, which must be submitted to the USEPA every four years. The lists also may include threatened waters.
2. Requiring states to prioritize water bodies and develop TMDLs first for those that are drinking water sources or support endangered species. Once a TMDL is developed, the rule requires states to establish a cleanup schedule that would enable polluted water bodies to achieve water quality standards within ten years (with fifteen years if the state requests and USEPA grants an extension).
3. An implementation plan that identifies specific actions and schedules for meeting water quality goals and addresses both point and non-point pollution sources, according to the rule. The rule also requires that runoff controls be installed five years after this plan is developed, if practicable, and that TMDL allocations for non-point sources be

   • pollution specific;
   • implemented expeditiously;
   • met through effective programs; and
   • supported by adequate water quality funding.

The rule does not require new permits for forestry, livestock, or aquaculture operations. It also does not require "offsets" for new pollution discharges to impaired waters prior to TMDL development.

DDT and Biomagnification

Figure 8.18 shows how DDT becomes concentrated in the tissues of organisms representing four successive trophic levels in a food chain. The concentration effect occurs because DDT is metabolized and excreted much more slowly than the nutrients that are passed from one trophic (feeding) level to the next. The result is that DDT accu-

TERTIARY CONSUMERS (13.8 ppm)

⇑

SECONDARY CONSUMERS (2.07 ppm)

⇑

PRIMARY CONSUMERS (0.23 ppm)

⇑

PRODUCERS (0.04 ppm)

Figure 8.18.   DDT and biomagnification: the numbers represent the value of the concentration in the tissues of DDT and its derivatives (in parts per million, ppm).

mulates in the bodies (especially in fat). Thus most of the DDT ingested as part of gross production is still present in the net production that remains at that trophic level.

✔ **Key Point:** This is why the hazard of DDT to nontarget animals is particularly acute for those species living at the top of food chains.

Chemistry and the Clean Water Act

The Clean Water Act (CWA) regulates the emission of hazardous pollutants into the nation's surface waters.

General pollutants entering surface waters include pH, toxic substances, suspended solids, oil and grease, pathogenic microorganisms, nutrients, nontoxic pollutants, and biochemical oxygen demand (BOD). (See chapter 15 for a discussion of BOD.)

The environmental practitioner responsible for maintaining water quality in surface waters is concerned with the fate of all these general pollutants. While some of the pollutants are biological, many are chemical. Thus, the environmental professional must be cognizant of and familiar with these chemicals, their sources, and their pertinent regulations. For example, regarding chemical contamination of our waterways, the environmental professional should be familiar with the following:

- What regulations affect surface water pollution? How have they affected the quality of surface water supplies?
- The usual surface water pollutants and their sources.
- Watersheds and non-point-source pollution.
- Problems associated with non-point-source pollution, and how those concerns are being handled.
- Why the EPA considers runoff the most serious water pollution problem in the U.S. Who will be affected by regulations controlling runoff, and how?
- Solid waste disposal in water systems.

Wastewater Treatment

Conventional wastewater treatment consists of primary and secondary treatment (see figure 8.19). Advanced, or tertiary, wastewater treatment consists of additional unit processes for nutrient removal.

✔ **Key Point:** Most wastewater treatment operations are operated by POTWs (Publicly Owned Treatment Works).

PRELIMINARY AND PRIMARY TREATMENT

The initial stage in the wastewater treatment process (following collection and influent pumping) is called simply *preliminary treatment*. Raw influent entering the treatment plant may contain many kinds of materials (trash). The purpose of preliminary treatment is to protect plant equipment by removing these materials that could cause clogs, jams, or excessive wear to plant machinery. In addition, the removal of various materials at the beginning of the treatment process saves valuable space within the treatment plant.

Preliminary treatment may include many different processes, each designed to remove a specific type of material that might be a potential problem for the treatment process. Processes include wastewater collection, influent pumping, screening, shredding, grit removal, flow measurement, preaeration, chemical addition, and flow equalization. The major processes are shown in figure 8.19.

The purpose of *screening* is to remove large solids such as rags, cans, rocks, branches, leaves, roots, etc. from the flow before the flow moves on to downstream processes.

The purpose of *grit removal* is to remove the heavy inorganic solids, which could cause excessive mechanical wear. Grit is heavier than inorganic solids and includes sand, gravel, clay, eggshells, coffee grounds, metal filings, seeds, and other similar materials.

The purpose of *primary treatment* (primary sedimentation or primary clarification) is to remove settleable organic and flotable solids. Normally, each primary clarification unit can be expected to remove 90% to 95% of settleable solids, 40% to 60%

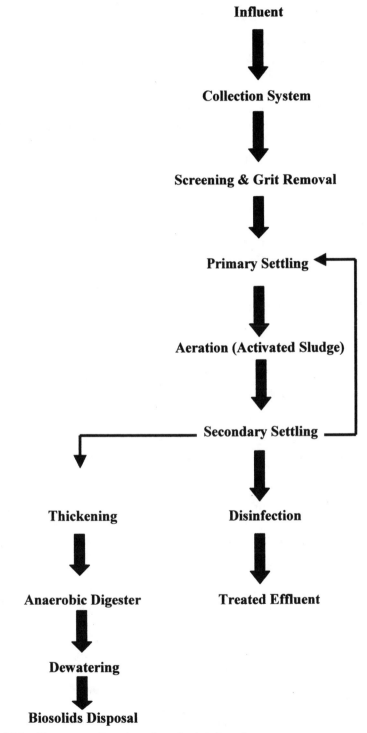

Figure 8.19.   The conventional wastewater treatment process.

of total suspended solids, and 25% to 35% of BOD_5 (biochemical oxygen demand; see below).

Sedimentation may be used throughout the plant to remove settleable and flotable solids. It is used in primary treatment, secondary treatment, and advanced wastewater treatment processes. In primary treatment or primary clarification, large basins are used in which primary settling is achieved under relatively quiescent conditions. Within these basins, mechanical scrapers collect the primary settled solids into a hopper, from which they are pumped to a sludge-processing area. Oil, grease, and other floating materials (scum) are skimmed from the surface. The effluent is discharged over weirs into a collection trough.

Upon completion of screening, degritting, and settling in sedimentation basins, large debris, grit, and many settleable materials have been removed from the waste stream. What is left is referred to as *primary effluent*. Usually cloudy and frequently gray in color, primary effluent still contains large amounts of dissolved food and other chemicals (nutrients). These nutrients are treated in the next step in the treatment process (secondary treatment).

SECONDARY TREATMENT

The main purpose of secondary treatment (sometimes referred to as biological treatment) is to provide *biochemical oxygen demand* (BOD) removal beyond what is achievable by primary treatment. There are three commonly used approaches, all of which take advantage of the ability of microorganisms to convert organic wastes (via biological treatment), into stabilized, low-energy compounds. Two of these approaches, the *trickling filter* (and/or its variation, the *rotating biological contactor* [RBC]) and the *activated sludge* process, sequentially follow normal primary treatment. The third, *ponds* (oxidation ponds or lagoons), however, can provide equivalent results without preliminary treatment.

Secondary treatment refers to those treatment processes that use biological processes to convert dissolved, suspended and colloidal organic wastes to more stable solids, which can be either removed by settling or discharged to the environment without causing harm. The Clean Water Act (CWA) defines secondary treatment as a process that produces an effluent with no more than 30 mg/L BOD_5 and 30 mg/L total suspended solids.

Most secondary treatment processes decompose solids aerobically producing carbon dioxide, stable solids, and more organisms. Since solids are produced, all of the biological processes must include some form of solids removal (settling tank, filter, etc.).

Secondary treatment processes can be separated into two large categories: fixed film systems and suspended growth systems.

Fixed film systems are processes that use a biological growth (biomass or slime), which is attached to some form of medium. Wastewater passes over or around the medium and the slime. When the wastewater and slime are in contact, the organisms remove and oxidize the organic solids. The medium may be stone, redwood, synthetic

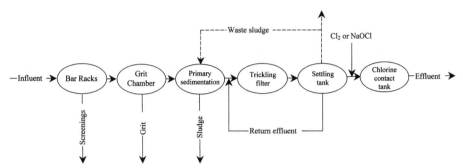

Figure 8.20.   Simplified flow diagram of a trickling filter.

materials, or any other substance that is durable (capable of withstanding weather conditions for many years), provides a large area for slime growth while providing open space for ventilation, and is not toxic to the organisms in the biomass. Fixed film devices include trickling filters and rotating biological contactors (RBCs).

Suspended growth systems are processes that use a biological growth that is mixed with the wastewater. Typical suspended growth systems consist of various modifications of the activated sludge process.

TREATMENT PONDS

Wastewater treatment can be accomplished using *ponds*. Ponds are relatively easy to build and manage, they accommodate large fluctuations in flow, and they can also provide treatment that approaches conventional systems (producing a highly purified effluent) at much lower cost. It is the cost (the economics) that drives many managers to decide on the pond option. The actual degree of treatment provided depends on the type and number of ponds used. Ponds can be used as the sole type of treatment or they can be used in conjunction with other forms of wastewater treatment; that is, other treatment processes followed by a pond or a pond followed by other treatment processes.

TRICKLING FILTERS

Trickling filters have been used to treat wastewater since the 1890s. It was found that if settled wastewater was passed over rock surfaces, slime grew on the rocks and the water became cleaner. Today public waterworks still use this principle but, in many installations, instead of rocks they use plastic media.

In most wastewater treatment systems, the *trickling filter* follows primary treatment and includes a secondary settling tank or clarifier, as shown in figure 8.20. Trickling filters are widely used for the treatment of domestic and industrial wastes. The process is a fixed film biological treatment method designed to remove BOD_5 and suspended solids.

A trickling filter consists of a rotating distribution arm that sprays and evenly distributes liquid wastewater over a circular bed of fist-sized rocks, other coarse materials, or synthetic media (see figure 8.21). The spaces between the medium's components allow air to circulate easily so that aerobic conditions can be maintained. The spaces also allow wastewater to trickle down through, around, and over the medium. A layer of biological slime that absorbs and consumes the wastes trickling through the bed covers the medium. The organisms aerobically decompose the solids, producing more organisms and stable wastes, which either become part of the slime or are discharged back into the wastewater flowing over the medium. This slime consists mainly of bacteria, but it may also include algae, protozoa, worms, snails, fungi, and insect larvae. The accumulating slime occasionally sloughs off individual medium materials and the *sloughings* are collected at the bottom of the filter, along with the treated wastewater, and passed on to the secondary settling tank, where it is removed.

ROTATING BIOLOGICAL CONTACTORS (RBCS)

The RBC is a biological treatment system (see figures 8.22–8.23) and a variation of the attached growth idea provided by the trickling filter. Still relying on microorganisms that grow on the surface of a medium, the RBC is instead a *fixed film* biological treatment device. The basic biological process, however, is similar to that occurring in the trickling filter. An RBC consists of a series of closely spaced (mounted side by side), circular, plastic (synthetic) disks that are typically about 3.5 m in diameter and attached to a rotating horizontal shaft. Approximately 40% of each disk is submerged in a tank containing the wastewater to be treated. As the RBC rotates, the attached biomass film (zoogleal slime) that grows on the surface of the disk moves into and out

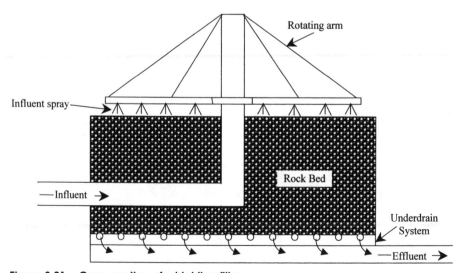

Figure 8.21.   Cross-section of a trickling filter.

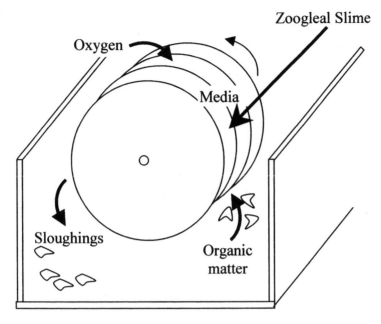

Figure 8.22.   Rotating biological contactor (RBC) cross-section.

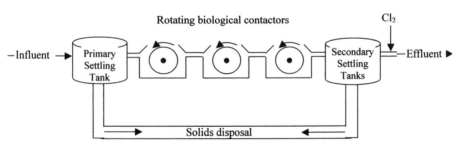

Figure 8.23.   Rotating biological contactor (RBC) treatment system.

of the wastewater. While submerged in the wastewater, the microorganisms absorb organics; while they are rotated out of the wastewater, they are supplied with needed oxygen for aerobic decomposition. As the zoogleal slime reenters the wastewater, excess solids and waste products are stripped off the media as sloughings. These sloughings are transported with the wastewater flow to a settling tank for removal.

ACTIVATED SLUDGE

The biological treatment systems discussed to this point (ponds, trickling filters, and RBCs) have been around for years. The trickling filter, for example, has been around and successfully used since the late 1800s. The problem with ponds, trickling filters and RBCs is that they are temperature sensitive and remove less BOD, and trickling filters, for example, cost more to build than the activated sludge systems that were later developed.

✔ **Note:** Although trickling filters and other systems cost more to build than activated sludge systems, it is important to point out that activated sludge systems cost more to operate because of the need for energy to run pumps and blowers.

As shown back in figure 8.19, the activated sludge (aeration) process follows primary settling. The basic components of an activated sludge sewage treatment system include an aeration tank and a secondary basin, settling basin, or clarifier (see figure 8.24). Primary effluent is mixed with settled solids recycled from the secondary clarifier and is then introduced into the aeration tank. Compressed air is injected continuously into the mixture through porous diffusers located at the bottom of the tank, usually along one side.

Wastewater is fed continuously into an aerated tank, where the microorganisms metabolize and biologically flocculate the organics. Microorganisms (activated sludge) are settled from the aerated mixed liquor under quiescent conditions in the final clarifier and are returned to the aeration tank. Left uncontrolled, the number of organisms would eventually become too great; therefore, some must periodically be removed (wasted). A portion of the concentrated solids (waste activated sludge or WAS from the bottom of the settling tank must be removed from the process). Clear supernatant from the final settling tank is the plant effluent.

DISINFECTION OF WASTEWATER

Like drinking water, liquid wastewater effluent is disinfected. Unlike drinking water, wastewater effluent is disinfected not to directly protect a drinking water supply (pipe-

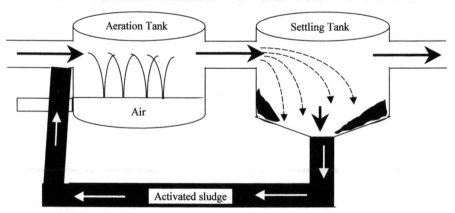

Figure 8.24. The activated sludge process.

to-pipe connection) but instead is treated to protect public health in general. This is particularly important when the secondary effluent is discharged into a body of water used for swimming or water supply for downstream consumption.

In the treatment of water for human consumption, treated water is typically chlorinated (although ozonation is also currently being applied in many cases). Chlorination is the preferred disinfection in potable water supplies because of chlorine's unique ability to provide a residual. This chlorine residual is important because, when treated water leaves the waterworks facility and enters the distribution system, the possibility of contamination is increased. The residual works to continuously disinfect water right up to the consumer's tap.

LAND APPLICATION

Land application is a way to process wastewater by applying secondary effluent onto a land surface. This procedure can provide an effective alternative to the expensive and complicated advanced treatment methods. A high-quality polished effluent (i.e., effluent in which levels of TSS, BOD, phosphorus, and nitrogen compounds as well as refractory organics are reduced) can be obtained by the natural processes that occur as the effluent flows over the vegetated ground surface and percolates through the soil.

Limitations are involved with land application of wastewater effluent. For example, the process needs large land areas. Soil type and climate are also critical factors in controlling the design and feasibility of a land treatment process.

TERTIARY TREATMENT (BNR)

Recent experience has shown that *biological nutrient removal* (BNR) systems are reliable and effective in removing nitrogen and phosphorus. The process is based upon the principle that, under specific conditions, microorganisms will remove more phosphorus and nitrogen than is required for biological activity. Several patented processes are available for this purpose. Performance depends on the biological activity and the process employed.

SOLIDS (SLUDGE/BIOSOLIDS)

The unit processes described to this point remove solids and BOD from the waste stream before the liquid effluent is discharged to its receiving waters. What remains to be disposed of is a mixture of solids and wastes, called *process residuals*—more commonly referred to as *sludge* or *biosolids*.

✔ **Note:** Sludge is the commonly accepted name for wastewater solids. However, if wastewater sludge is used for beneficial reuse (e.g., as a soil amendment or fertilizer), it is commonly called *biosolids*.

The most costly and complex aspect of wastewater treatment can be the collection, processing, and disposal of sludge. This is so because the quantity of sludge produced may be as high as 2% of the original volume of wastewater, depending somewhat on the treatment process being used.

Because sludge can be as much as 97% water, and because the cost of disposal will be related to the volume of sludge being processed, one of the primary purposes or goals (along with stabilizing it so it is no longer objectionable or environmentally damaging) of sludge treatment is to separate as much of the water from the solids as possible. Sludge treatment methods may be designed to accomplish both of these purposes.

When we speak of sludge or biosolids, we are speaking of the same substance or material; each is defined as the suspended solids removed from wastewater during sedimentation, then concentrated for further treatment and disposal or reuse. The difference between the terms *sludge* and *biosolids* is determined by the way they are managed. (Note: The task of disposing, treating or reusing wastewater solids is called *sludge* or *biosolids management.*) Sludge is typically seen as wastewater solids that are "disposed" of. Biosolids is the same substance managed for reuse—commonly called *beneficial reuse* (e.g., for land application as a soil amendment, such as biosolids compost).

✔ **Key Point:** As wastewater treatment standards have become more stringent because of increasing environmental regulations, the volume of wastewater sludge has increased.

✔ **Key Point:** Before sludge can be disposed of or reused, it requires some form of treatment to reduce its volume, to stabilize it, and to inactivate pathogenic organisms.

Sludge forms initially as a 3% to7% suspension of solids, and with each person typically generating about 4 gallons of sludge per week, the total quantity generated each day, week, month, and year is significant. Because of the volume and nature of the material, sludge management is a major factor in the design and operation of all water pollution control plants.

✔ **Note:** Wastewater solids account for more than half of the total costs in a typical secondary treatment plant.

CHEMICAL TREATMENT AND SOLIDS

The addition of chemicals and various organic and inorganic substances prior to sedimentation and clarification may increase the solids captured and reduce the amount of solids lost in the effluent. This *chemical addition* results in the formation of heavier solids, which trap the colloidal solids or convert dissolved solids to settleable solids. The resultant solids are known as *chemical sludges.* As chemical usage increases, so does the quantity of sludge that must be handled and disposed of. Chemical sludges can be very difficult to process; they do not de-water well and contain lower percentages of solids.

Additional Reading

Evangelou, V.P., *Environmental Soil and Water Chemistry: Principles and Applications.* New York: John Wiley & Sons, Inc., 1998.
Frei, R. W. W., et al., *Analysis and Chemistry of Water Pollution.* Vol. 6. Gordon & Breach Publishing Group, 1983.
Hrubec, Juri, ed., *Water Pollution: Drinking Water and Drinking Water Treatment.* New York: Springer-Verlag Inc., 1995.
Spellman, F. R., *Handbook of Water and Wastewater Treatment Plant Operations.* New York: Lewis Publishers, 2003.

Summary

- Water is essential to life.
- Water is transparent.
- Water is an excellent solvent.
- Water is a polar solvent.
- Water has high surface tension.
- Water has its maximum liquid density at 4°C.
- Hydrogen bonding increases the boiling point of water above that predicted based on molecular weight.
- Water has a higher heat capacity than any other liquid except ammonia.
- Water has a higher heat of vaporization than any other material.
- Water has a higher latent heat of fusion than any other liquid except ammonia.
- A solution is a homogeneous mixture of a solute in a solvent.
- The solvent can be separated from its solution by distillation.
- Polar solutes dissolve in polar solvents.
- Nonpolar solutes dissolve in nonpolar solvents.
- A concentrated solution has a large amount of solute dissolved in the solution.
- A dilute solution has a small amount of solute dissolved in its solution.
- The acidity of natural water systems is defined as its capacity to neutralize OH^-.
- Alkalinity is defined as the capacity of water to accept H^+.
- Alkalinity is usually expressed by chemists in terms of equivalents per liter.
- The salinity of water is its salt load.
- Water hardness is attributed to the concentration of Ca^{+2}.
- Common aqueous metal contaminants include common metals, heavy metals, metalloids, organometallics, and radionuclides.
- Organic and inorganic substances contribute to water pollution.
- TMDL is the amount of a pollutant that can be discharged to a water body and still attain water quality goals.

- Biomagnification is demonstrated when DDT becomes concentrated as it passes through a food chain.
- Primary treatment is the first step in the wastewater treatment process.
- Secondary wastewater treatment uses biochemical processes to digest organic wastes.
- Secondary treatment uses trickling filters, rotating biological contactors, and activated sludge.
- Tertiary treatment involves one or more physical, chemical, and/or biochemical processes used to remove nutrients.

New Word Review

Activated sludge—the solids formed when microorganisms are used to treat wastewater using the activated sludge treatment process. It includes organisms, accumulated food materials, and waste products from the aerobic decomposition process.

Aerobic—said of conditions in which free, elemental oxygen is present. Also used to describe organisms, biological activity, or treatment processes that require free oxygen.

Alkalinity—a measure of water's capacity to absorb hydrogen ions without significant pH change (i.e., to neutralize acids).

Anaerobic—said of conditions in which no oxygen (free or combined) is available. Also used to describe organisms, biological activity, or treatment processes that function in the absence of oxygen.

Colligative properties—properties of a solution that depend only on the concentration of a solute species.

Colloidal—said of any substance in a certain state of fine division in which the particles are less than one micron in diameter.

Compound—a substance of two or more chemical elements chemically combined. Example: water (H_2O) is a compound formed by hydrogen and oxygen.

Concentration—how much solute (what is being dissolved) is contained in a solution (what contains the solute).

Grit—heavy inorganic solids, such as sand, gravel, eggshells, or metal filings.

Influent—the wastewater entering a tank, channel, or treatment process.

Inorganic—chemical substances of mineral origin.

Liquids—have a definite volume, but not shape that will fill containers to certain levels and form free level surfaces.

Mixture—a physical, not chemical, intermingling of two or more substances. Example: sand and salt stirred together.

Nutrient—substance required to support living organisms. Usually refers to nitrogen, phosphorus, iron, and other trace metals.

Organic—chemical substance of animal or vegetable origin, made, basically, of carbon structure.

Precipitate—a solid substance that can be dissolved but is separated from solution because of a chemical reaction or change in conditions, such as pH or temperature.

Saturated solution—the physical state in which a solution will no longer dissolve more of the dissolving substance—solute.

Sewage—wastewater-containing human wastes.

Sludge—the mixture of settleable solids and water that is removed from the bottom of the settling tank.

Solids—substances that maintain definite size and shape.

Solute—the component of a solution that is dissolved by the solvent.

Solvent—the component of a solution that does the dissolving.

Titration—the method of determining the concentration of a solution by adding a solution of a reactant to a solution of sample until an indicator changes color.

Wastewater—the water supply of the community after it has been soiled by use.

Chapter Review Questions

8.1. The chemical symbol for sodium is _____.

8.2. The chemical symbol for sulfuric acid is _____.

8.3. Neutrality on the pH scale is _____.

8.4. The chemical symbol for calcium carbonate is _____.

8.5. Is NaOH a salt or a base?

8.6. Chemistry is the study of substances and the _____ they undergo.

8.7. The three states of matter are _____, _____ and _____.

8.8. A basic substance that cannot be broken down any further without changing the nature of the substance is an _____.

8.9. A combination of two or more elements is a _____.

8.10. A table of the basic elements is called the _____ table.

8.11. When a substance is mixed into water to form a solution, the water is called the _____ and the substance is called the _____.

8.12. Define ion.

8.13. A solid that is less than 1 micron in size is called a _____.

8.14. The property of water that causes light to be scattered and absorbed is _____.

8.15. What is the main problem with metals found in water?

8.16. Compounds derived from material that once was alive are called _____ chemicals.

8.17. pH range is from _____ to _____.

8.18. What is alkalinity?

8.19. The two ions that cause hardness are _____ and _____.

8.20. What type of substances produces hydroxide ions (OH^-) in water?

Notes

1. F. R. Spellman and J. E. Drinan, *Stream Ecology & Self-Purification: An Introduction*, 2nd ed. (Lancaster, Pa.: Technomic, 2001), 133–34.

2. Spellman and Drinan, *Stream Ecology*, 135–37.

CHAPTER 9

Atmospheric Chemistry

> This we know: All things are connected like the blood that unites us. We did not weave the web of life; we are merely a strand in it. Whatever we do to the web, we do to ourselves.
>
> We love this earth as a newborn loves its mother's heartbeat. If we sell you our land, care for it as we have cared for it. Hold in your mind the memory of the land as it is when you receive it.
>
> Preserve the land and the air and the rivers for your children's children and love it as we have loved it.
>
> —Chief Seattle, mid-1850s

Topics in This Chapter

- Earth's Atmosphere: "A Flask without Walls"
- Functions of the Atmosphere
- Structure of the Atmosphere
- Chemical Reactions in the Atmosphere
- Air Pollutants
- Chemistry and the Clean Air Act (CAA)

Earth's Atmosphere: "A Flask without Walls"

In the following discussion of the chemistry of Earth's atmosphere, you should bear in mind that I refer to the atmosphere as it is at present (during the age of humans). The atmosphere was chemically quite different in previous eras. Note also that "atmospheric chemistry" is a scientific discipline that can stand on its own. An in-depth presentation of atmospheric chemistry is beyond the scope of this book. Here, certain important atmospheric chemistry phenomena are highlighted, especially those problems caused by organic and inorganic air pollutants.

The full range of chemistry all occurs in the atmosphere—the atmosphere is a "flask without walls."[1] Excluding highly variable amounts of water vapor, more than 99% of the molecules constituting Earth's atmosphere are nitrogen, oxygen, and chemically inert gases (noble gases such as argon, etc.).

The chemistry, and thus the reactivity, of these natural gases (nitrogen, oxygen, carbon dioxide, argon, and others) is well known. The other reactive chemicals (anthropogenic, or produced by humans) that are part of Earth's atmosphere are also known, but there are still differing opinions on their exact total effect on our environment. For example, methane is by far the most abundant reactive compound in the atmosphere, and it currently is at a ground-level concentration (in the Northern Hemisphere) of about 1.7 ppmv. We know significant amounts of information about methane (its generation and fate when discharged) and its influence on the atmosphere; however, we are still conducting research to find out more, as we should.

Many different reactive molecules (other than methane) exist in the atmosphere. We may not be familiar with each of these reactants, but many of us certainly are familiar with their consequences: the ozone hole, the greenhouse effect and global warming, smog, acid rain, the rising tide, and so on. It may surprise you to know, however, that the total amount of all these reactants in the atmosphere is seldom more than 10 ppmv anywhere in the world at any given time. The significance should be obvious: the atmospheric problems currently occurring on Earth are the result of less than one thousandth of 1% of all the molecules in the atmosphere. This indicates that environmental damage (causing global atmospheric problems) can result from far less than the tremendous amounts of reactive substances we might imagine are dangerous.

Various contaminants released to the atmosphere manifest a variety of environmental problems. Consider, for example, the following airborne contaminants and their implications in global warming, acid rain, distortion of visibility, increased respiratory problems, and/or plant necrosis (yellowing):

- Particulates—distort visibility, increase respiratory problems
- CO—greenhouse contributor because it slowly breaks down into CO_2
- CH_4—greenhouse gas
- Fly ash—distorts visibility, increases respiratory problems
- CFCs—greenhouse gas
- N_2O—greenhouse gas, acid rain, increased respiratory problems, plant necrosis
- CO_2—greenhouse gas
- SO_2—acid rain, increased respiratory problems, plant necrosis
- NO_x—greenhouse gas, acid rain, increased respiratory problems, plant necrosis
- Coal dust—distorts visibility, increases respiratory problems

More specifically, various chemical activities contribute to environmental pollution. In addition to the implications listed above, the types of environmental pollution and sources by the chemical industry include sulfur dioxide, toxic gas emissions, foul-smelling gases, dust, smoke, sprays, and radioactivity.

The quality of the air we breathe, visibility and atmospheric esthetics, and our climate are important to our health and to our quality of life, and all are dependent upon chemical phenomena that occur in the atmosphere. Global atmospheric problems, such as the nature and level of air pollutants, concern the environmental practitioner the most because they affect health, quality of life, and the environment.

Functions of the Atmosphere

Earth's atmosphere has many functions:

- It is a reservoir of gases of use to the biosphere carbon dioxide (photosynthesis), oxygen (respiration), and nitrogen (nitrogen fixation), and water.
- It is a protective shield for the biosphere (role of ozone).
- It is a transport medium for energy and water.
- It is a medium for waste disposal.
- It moderates temperature (transports heat).

✔ **Important Point:** The atmosphere has a large capacity to absorb, dilute, and remove pollutants. Its capacity to dilute and disperse varies by a factor of ~10,000, depending on factors such as atmospheric stability, but is *not limitless.*

Structure of the Atmosphere

Simply defined, the atmosphere is an envelope of gases and particles that surround Earth. The atmosphere is characterized as a compressible fluid, is held in place by gravity, and is a mixture of gases (some condensable) and particles.

Earth's atmosphere is "paper" thin—approximately 1/1,000,000 the mass of the Earth. Shakespeare likened it to a majestic overhanging roof; others have seen it as an envelope, a veil, or a gaseous shroud. However, the atmosphere is more like the skin of an apple. This thin skin, or layer, contains the life-sustaining oxygen (21%) required by all humans and many other life-forms; the carbon dioxide (0.03%) so essential for plant growth; the nitrogen (78%) needed for chemical conversion to plant nutrients; the trace gases (all free elements) such as methane, argon, helium, krypton, neon, xenon, ozone, and hydrogen; and varying amounts of water vapor and airborne particulate matter.

✔ **Key Point:** It is interesting to note that whenever we take a breath of fresh air we are likely to inhale a mixture of many free elements, including N, O, H, He, Ar, Ne, Kr, Xe, and possibly Rn.

Gravity holds about half the weight of a fairly uniform mixture of these gases in the lower 18,000 feet of the atmosphere; 98% of the material in the atmosphere is below 100,000 feet.

More than 99% of the atmospheric mass is found within 40 km of Earth's surface. The upper atmosphere is very sparse.

From figure 9.1 we see that the atmosphere has a thickness of 40–50 miles. Figure 9.1 also shows that the gaseous area surrounding Earth is divided into several concentric spherical strata separated by narrow transition zones.

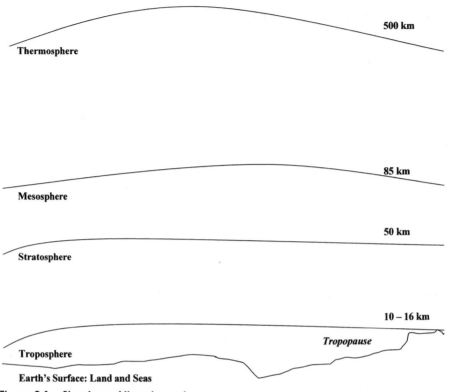

Figure 9.1. Structure of the atmosphere.

TROPOSPHERE

The troposphere is the atmospheric layer closest to Earth and contains the largest percentage of the mass of the total atmosphere. The density of its air and an average vertical temperature change of 6° Celsius (C) per kilometer characterize it.

The troposphere is where all the weather takes place; it is the region of rising and falling packets of air. The air pressure at the top of the troposphere and the next layer is called the *tropopause*.

✔ **Key Point:** The sun produces an enormous amount of energy that would have vaporized Earth long ago except for the atmosphere's mitigating effects (see figure 9.2).

STRATOSPHERE

Above the troposphere is the stratosphere, where airflow is mostly horizontal. The thin *ozone layer* in the upper stratosphere has a high concentration of ozone, a particularly reactive form of oxygen.

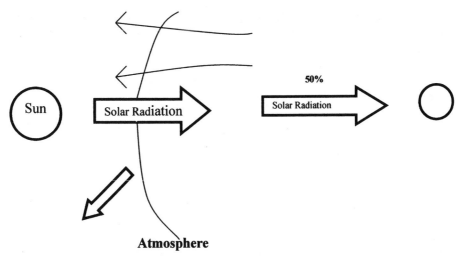

Figure 9.2. The atmosphere's mitigating effect against solar radiation.

The ozone layer (life-protecting ozone should not be confused with pollutant ozone), which is produced by the intense ultraviolet radiation from the sun, is small in quantity—if it were compressed to a liquid layer over the globe at sea level, it would have a thickness of less than three sixteenths of an inch—is critical to our survival. Ozone plays an important role in regulating the thermal regime of the stratosphere, as water vapor content within the layer is very low.

🗸 **Key Point:** There is considerable recent concern that human-made chlorofluoro-carbon compounds (CFCs) may be depleting the ozone layer, with dire future consequences for life on Earth.

🗸 **Key Point:** Earth's heat balance is maintained by a complex series of factors that are not well understood. Small variations in temperature can have significant effects on Earth's climate.

THERMOSPHERE

The thermosphere is located above the stratosphere and is separated from it by a transition layer (*mesosphere*). The temperature in the thermosphere generally increases with altitude up to 1,000–1,500 Kelvin. This increase in temperature is due to the absorption of intense solar radiation by the limited amount of remaining molecular oxygen. At an altitude of 100–200 km, the major atmospheric components are still nitrogen and oxygen. At this extreme altitude, gas molecules are widely separated.

EARTH'S HEAT BALANCE

Approximately 50% of the solar radiation entering the atmosphere reaches Earth's surface, either directly or after being scattered by clouds, particulate matter, or atmospheric gases. The other 50% is either reflected directly back or absorbed in the atmosphere, and its energy is reradiated back into space at a later time as infrared radiation. Most of the solar energy reaching the surface is absorbed and must be returned to space to maintain heat balance.

Energy is transferred or transported in our atmosphere by three mechanisms (see figure 9.3):

- Conduction (adjacent molecules interacting)
- Convection (bulk movement)
- Radiation (infrared)

Conduction of energy occurs through the interaction of adjacent molecules with no visible motion accompanying the transfer of heat; for example, the whole length of a metal rod will become hot when one end is held in a fire. Because air is a poor heat conductor, conduction is restricted to the layer of air in direct contact with Earth's surface.

The heated air is then transferred aloft by convection, the movement of whole masses of air, which may be either relatively warm or cold. *Convection* is the mechanism by which abrupt temperature variations occur when large masses of air move across an area. Air temperature tends to be higher near the surface of Earth and decreases gradually with altitude. A large amount of Earth's surface heat is transported to clouds in the atmosphere by conduction and convection before being lost ultimately

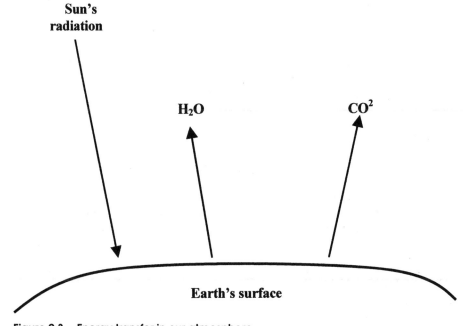

Figure 9.3. Energy transfer in our atmosphere.

by radiation, and this redistribution of heat energy plays an important role in weather and climate conditions.

Radiation of energy occurs through electromagnetic radiation in the infrared region of the spectrum. The crucial importance of the radiation mechanism is that it carries energy away from Earth on a wavelength much longer than that of the solar energy (sunlight) that brings energy to Earth and, in turn, works to maintain Earth's heat balance.

Chemical Reactions in the Atmosphere

Atmospheric reactions generally occur at very low pressures (i.e., low concentrations) in the presence of large amounts of energy (i.e., sunlight).

✔ **Interesting Point:** Manahan points out that the "study of atmospheric chemical reactions is difficult." Why? "Because the chemist must deal with incredibly low concentrations, so that detection and analysis of reaction products is quite difficult. It is difficult to simulate these reactions in the laboratory because even the walls of the container can act as a "third body" to absorb energy or act as a catalyst."[2]

Because of their low concentrations of atoms and molecules, chemically reactive species tend to persist in the upper atmosphere. The most important reactive species are:

- Photochemically excited species X*
- Ions and electrons (charges species)
- Free radicals

✔ **Key Point:** An asterisk (*) designates an electronically excited molecule.

Not surprisingly, *photochemical reactions*, which occur in the presence of light, play a significant role in atmospheric chemistry.

✔ **Key Point:** Many of these reactions would not occur in the absence of light, especially under the conditions (e.g., temperatures) found in the upper atmosphere.

EXCITED STATE X*

Chemicals in the atmosphere can absorb light to form an *excited state X**. These excited molecules primarily lose energy by dissociation, photoionization, and direct reaction. Dissociation is a bond broken by absorption of a photon (light):

$$O_2^* \rightarrow O + O$$

Photoionization—removal of an outer (valence) electron from a molecule by absorption of a photon:

$$N_2^* \rightarrow N_2^* + e^-$$

Direct reaction:

$$O_2^* + O_3 \rightarrow 2O_2 + O$$

IONS AND ELECTRONS

Ions and electrons (i.e., charged atoms or molecular fragments) are present in such high quantities in the upper atmosphere (> 50 km; because solar radiation is very intense) that this region is called the *ionosphere*. Radio transmission around the curvature of Earth is possible because radio waves bounce off the ionosphere.

✔ **Interesting Point:** At night, the formation of ions ceases and the lower limit of the ionosphere "lifts" because of recombination, allowing radio wave transmissions over greater distances.

FREE RADICALS

Free radicals are composed of atoms or molecular fragments with unshared electrons. They are formed by the action of high-energy radiation (i.e., sunlight; see figure 9.4).

These radicals can react with other species to form new radicals or be "quenched" by another radical. The *hydroxyl radical,* an extremely important chemical species in atmospheric chemistry, is produced by many reactions, including:

$$H_2 \rightarrow + h\nu \; HO^* + H$$
$$O_3 \rightarrow + h\nu \; O^* + O_2$$
$$O^* + H_2O \rightarrow h\nu \; 2HO^*$$

REACTIONS OF ATMOSPHERIC OXYGEN

Oxygen (O_2—Greek *oxys,* "acid"; *genes,* "forming") constitutes approximately a fifth (21% by volume and 23.2% by weight) of the air in Earth's atmosphere. Gaseous

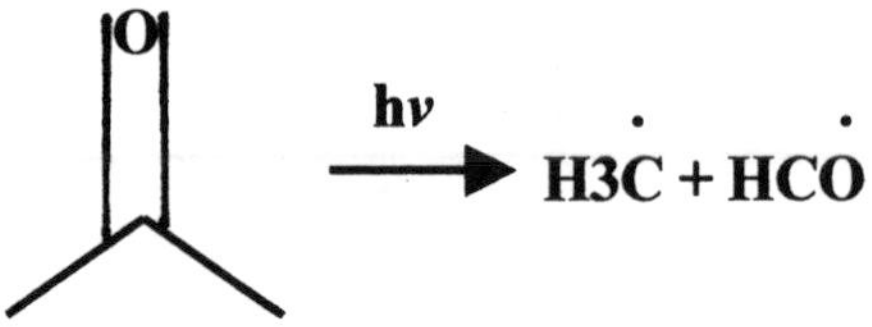

Figure 9.4. Free radicals formed by the action of high-energy radiation.

oxygen (O_2) is vital to life as we know it. On Earth, oxygen is the most abundant element. Most oxygen on Earth is not found in the free state but in combination with other elements as chemical compounds. Water and carbon dioxide are common examples of compounds that contain oxygen, but there are countless others.

Photosynthetic organisms are thought to have generated all the *molecular* oxygen (O_2) in the atmosphere. In the upper atmosphere, *elemental* oxygen and other forms also exist: O, O^*, O_2^*, and ozone (O_3).

✔ **Key Point:** Less than 10% of the oxygen at altitudes > 400 km exists in the form of O_2.

The physical properties of oxygen (O_3) are noted in table 9.1.

Atomic oxygen (O) is produced by the photochemical decomposition of molecular oxygen. It is found in the rarified *thermosphere* (see figure 9.1).

$$O_2 + hv \rightarrow O + O$$

O^+ is the principal cation in parts of the ionosphere.

$$O + hv \rightarrow O^+ + e^-$$

Ozone (O_3)—just another form of oxygen—is a highly reactive pale blue gas with a penetrating odor. Ozone is an *allotropic* modification of oxygen. An allotrope is a variation of an element that possesses a set of physical and chemical properties significantly different from the normal form of the element. Only a few elements have allotropic forms: oxygen, phosphorus, and sulfur are a few of them. Formed when ultraviolet (UV) radiation or electrical discharge splits the molecule of the stable form of oxygen (O2), it has three, instead of two, atoms of oxygen per molecule. Thus, its chemical formula is represented by O_3.

Ozone forms a thin layer (concentrations as high as 10%) in the upper atmosphere and serves as a radiation shield (230–330 nm) that protects life on Earth from ultraviolet rays (a cause of skin cancer). At lower atmospheric levels, it is an air pollutant and contributes to the greenhouse effect. At ground level, ozone, when inhaled,

Table 9.1.   Physical properties of oxygen.

Chemical formula	O_2
Molecular weight	31.9988
Freezing point	$-361.12°F$
Boiling point	$-297.33°F$
Heat of fusion	5.96 Btu/lb
Heat of vaporization	91.70 Btu/lb
Density of gas at boiling point	0.268 lb/ft³
Density of gas at room temperature	0.081 lb/ft³
Vapor density (air = 1)	1.105
Liquid-to-gas expansion ratio	8.75

can cause asthma attacks, stunted growth in plants, and corrosion of certain materials. Produced by the action of sunlight on air pollutants (including car exhaust fumes), ozone is a major air pollutant in hot summers.

$$O_2 + h\nu \rightarrow O + O \ (<242.4 \ nm)$$
$$O + O_2 + M \rightarrow O_3 \ M \ (increased \ energy)$$

✔ **Key Point:** The radiation absorbed by O_3 is transformed into heat.

✔ **Key Point:** Ozone is a toxic pollutant in the troposphere.

REACTIONS OF ATMOSPHERIC NITROGEN

Nitrogen (N_2) makes up the major portion of the atmosphere (78.03% by volume, 75.5% by weight). A colorless, odorless, tasteless, nontoxic, and almost totally inert gas, nitrogen is nonflammable, will not support combustion, and is not life supporting.

Nitrogen is part of Earth's atmosphere primarily because, over time, it has simply accumulated in the atmosphere and remained in place and in balance. This nitrogen accumulation process has occurred because, chemically, nitrogen is not very reactive. When released by any process, it tends not to recombine with other elements and accumulates in the atmosphere.

✔ **Key Point:** We need nitrogen not for breathing, but for other life-sustaining processes.

Although nitrogen in its gaseous form is of little use to us, after oxygen, carbon, and hydrogen, it is the most common element in living tissues. As a chief constituent of chlorophyll and amino acids and nucleic acids—the "building blocks" of proteins (which are used as structural component in cells)—nitrogen is essential to life. Animals cannot use nitrogen directly but only by way of the plant or animal tissues they eat. Plants obtain the nitrogen they need in the form of inorganic compounds, principally nitrate and ammonium.

Gaseous nitrogen is converted to a form usable by plants (nitrate ions) chiefly through the process of nitrogen fixation via the nitrogen cycle, shown in simplified form in figure 9.5.

✔ **Key Point:** UV light at altitudes below 100 km does not readily dissociate diatomic nitrogen.

$$N_2 + h\nu \rightarrow N + N$$

The physical properties of nitrogen are noted in table 9.2.

Nitrogen oxides—usually collectively symbolized by the formula NO_x—include nitrous oxide (N_2O), nitric oxide (NO), dinitrogen trioxide (N_2O_3), nitrogen dioxide (NO_2), dinitrogen tetroxide (N_2O_4), and dinitrogen pentoxide (N_2O_5).

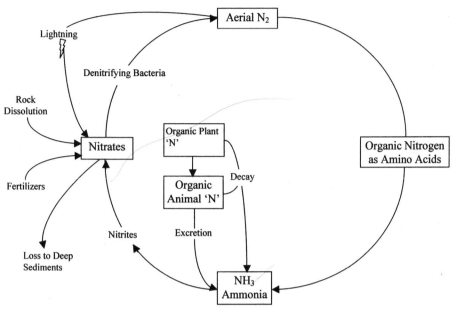

Figure 9.5.   Nitrogen cycle.

Table 9.2.   Physical properties of nitrogen.

Chemical formula	N_2
Molecular weight	28.01
Density of gas at 70°F	0.072 lb/ft³
Specific gravity of gas at 70°F and 1 atm (air = 1)	0.967
Specific volume of gas at 70°F and 1 atm	13.89 ft³/lb
Boiling point at 1 atm	−320.4°F
Melting point at 1 atm	−345.8°F
Critical temperature	−232.4°F
Critical pressure	493 psia
Critical density	19.60 lb/ft³
Latent heat of vaporization at boiling point	85.6 Btu/lb
Latent heat of fusion at melting point	11.1 Btu/lb

Nitric oxide, nitrogen dioxide, and nitrogen tetroxide are so-called fire gases. One or more of them is generated when certain nitrogenous organic compounds (polyurethanes) burn. Nitric oxide is the product of incomplete combustion, and a mixture of nitrogen dioxide and nitrogen tetroxide is the product of complete combustion.

Nitric oxide (NO) is thought to be the primary mechanism by which O_3 is removed from the stratosphere.

$$O_3 + NO \rightarrow NO_2 + O_2$$
$$NO_2 + O \rightarrow NO + O_2 \text{ (regeneration of NO)}$$

✔ **Key Point:** NO_2 is a primary cause of photochemical smog.

$$NO_2 + h\nu \rightarrow NO + O$$

ATMOSPHERIC CARBON DIOXIDE

Carbon dioxide (CO_2) is a colorless, odorless gas (although some people feel it has a slight pungent odor and biting taste), slightly soluble in water, more dense than air (one and a half times heavier than air), and slightly acidic. Carbon dioxide gas is relatively nonreactive and nontoxic. It will not burn, and it will not support combustion or life.

CO_2 is normally present in atmospheric air at about 0.035% by volume and cycles through the biosphere (carbon cycle), as shown in figure 9.6. Carbon dioxide, along with water vapor, is primarily responsible for the absorption of infrared energy re-emitted by Earth. In turn, some of this energy is reradiated back to Earth's surface. It is also a normal end product of human and animal metabolism. Our exhaled breath contains up to 5.6% carbon dioxide. Burning carbon-laden fossil fuels also releases carbon dioxide into the atmosphere. Much of this carbon dioxide is absorbed by ocean water; some of it is taken up by vegetation through photosynthesis in the carbon cycle, and some remains in the atmosphere.

✔ Interesting Point: Today, scientists estimate that the concentration of carbon dioxide in the atmosphere is approximately 350 ppm and increases at a rate of approximately 20 ppm every decade. The increasing rate of combustion of coal and oil has been primarily responsible for this occurrence, which may eventually have an impact on global climate.

The physical properties of carbon dioxide are noted in table 9.3.

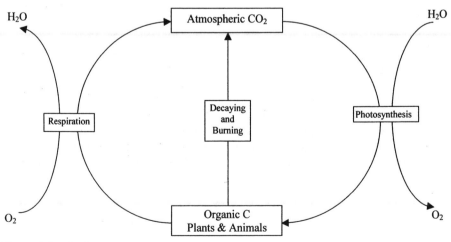

Figure 9.6. Carbon cycle.

Table 9.3.   Physical properties of carbon dioxide.

Chemical formula	CO_2
Molecular weight	44.01
Vapor pressure at 70°F	838 psig
Density of gas at 70°F and 1 atm	0.1144 lb/ft³
Specific gravity of gas at 70°F and 1 atm (air = 1)	1.522
Specific volume of gas at 70°F and 1 atm	8.741 ft³/lb
Critical temperature	− 109.3°F
Critical pressure	1070.6 psia
Critical density	29.2 lb/ft³
Latent heat of vaporization at boiling point	100.8 Btu/lb
Latent heat of fusion at − 69.9°	85.6 Btu/lb

Air Pollutants

Traditionally, it has been convenient to categorize air pollutants as being either organic (particulates, carbon oxides, sulfur dioxide, and nitrogen oxides) or inorganic (direct pollutants that cause cancer, etc., and secondary pollutants that contribute to photo-chemical smog). Even though Manahan[3] points out that "there is a strong connection between inorganic and organic substances in the atmosphere," I prefer to stick to tradition, discussing various air pollutants separately, as inorganic and organic pollutants.

Before discussing various air pollutants, however, I would emphasize that, strictly speaking, too much of any substance in the wrong place or at the wrong time is a pollutant. More pointedly, atmospheric pollution may be defined as "the presence of substances in the atmosphere, resulting from human-made activities or from natural processes, causing adverse effects to man and the environment. . . . [A]ir pollution is a term used to describe any unwanted chemicals or other materials that contaminate the air that we breathe resulting in the degradation of air quality."[4]

✔ **Key Point:** The inorganic chemicals discussed below are all typical urban air pollutants from human-made activities—nitrogen oxides, carbon monoxide, sulfur dioxide, hydrocarbons, and particulates. All these pollutants are called *primary pollutants* because they are emitted directly into the atmosphere.

INORGANIC AIR POLLUTANTS

Particulate Matter

Hastie points out that *atmospheric particulate matter* (aerosols, or PM) is important in the atmosphere for at least four reasons: "It provides a sink for reactive gases; e.g., particulate matter contains sulphate and nitrate from oxidation of SO_2 and NO_2. It provides a surface for chemical reactions; the chemistry of Antarctic ozone depletion attests to that. Particles can carry toxic chemicals such as PAHs and metals to remote areas; many of the toxic chemicals found in the Arctic get there this way. Particles can penetrate into the bronchial tract and the lungs causing human health problems such as bronchitis and pneumonia; there is now overwhelming evidence relating mortality to particulate levels."[5]

Atmospheric particulate matter is defined to be any dispersed matter, solid or liquid, in which the individual aggregates are larger than single small molecules (about 0.0002 μm in diameter), but smaller than about 500 μm.[6] In reality, particulate matter is a collective term used to describe small solid and liquid particles that are present in the atmosphere over relatively brief (minutes) to extended (days to weeks) periods of time. Individual particles vary in size, geometry, mass, concentration, chemical composition, and physical properties. They may be produced naturally or as a direct or indirect result of human activities.

Significant numbers of particulate matter are suspended in the atmosphere, particularly in the troposphere. These particles originate in nature from smokes, sea sprays, dusts, and the evaporation of organic materials from vegetation. A wide variety of nature's living or semi-living particles are also present—spores and pollen grains, mites and other tiny insects, and diatoms. The atmosphere also contains a bewildering variety of anthropogenic (man-made) particles produced by automobiles, refineries, production mills, and many other human activities. Elements in particulate matter include:

- Al, Fe, Ca, Si from soil, rocks, or coal combustion
- C as a result of incomplete combustion
- Na, Cl from marine aerosols, incineration of organohalide polymers
- Sb, Se from combustion of oil, coal, or refuse
- V, Zn, Pb resulting from combustion
- Be, Ca, Cr, Ni, As, Hg from various sources

Atmospheric particulate matter varies greatly in size, ranging over many orders of magnitude from molecular clusters to those that are visible with the unaided eye. Colloidal-sized particles in the atmosphere are called *aerosols*—usually less than 0.1 mm in diameter; the smallest are gaseous clusters and ions and submicroscopic liquids and solids; somewhat larger ones produce the beautiful blue haze in distant vistas; those two to three times larger are highly effective in scattering light; and the largest consist of such things as rock fragments, salt crystals, and ashy residues from volcanoes, forest fires, or incineration.

The largest number of airborne particulates is always in the invisible range. These numbers vary from less than one per liter to more than a half million per cubic centimeter in heavily polluted air and to at least ten times more than that when a gas-to-particle reaction is occurring.[7]

Based on particulate level, two distinct kinds of regions appear in the atmosphere: the very clean, and the dirty. The clean parts hold so few particulates that they are almost invisible, making them hard to collect or measure. In the dirty parts of the atmosphere—the air of a large metropolitan area—the concentration of particles includes an incredible quantity and variety of particulates from a wide range of sources.

Atmospheric particulate matter performs a number of functions, undergoes several processes, and is involved in many chemical reactions in the atmosphere. Probably the most important function of particulate matter in the atmosphere is its action as nuclei for the formation of water droplets and ice crystals. Much of the work of Vincent J. Schaefer (inventor of cloud seeding) involved using dry ice in early attempts, but it later evolved around the addition of condensing particles to atmospheres supersaturated with water vapor and the use of silver iodide, which forms huge numbers of very small particles. Another important function of atmospheric particulate matter is that it helps determine the heat balance of Earth's atmosphere by reflecting light. Particulate matter is also involved in many chemical reactions in the atmosphere—neutralization, catalytic effects, and oxidation reactions.

Carbon Oxides

Carbon monoxide (CO) is a colorless, odorless, tasteless gas that is by far the most abundant of the primary pollutants, as table 9.4 indicates.

Carbon monoxide has little direct effect on ecosystems but has an indirect environmental impact by contributing to the greenhouse effect and depletion of Earth's protective ozone layer.

✔ **Important Point:** The residence time of CO in the atmosphere is approximately four months.

The most important natural source of atmospheric carbon monoxide is the combination of oxygen with methane (CH_4), a product of the anaerobic decay of vegetation. ("Anaerobic" means that the decay occurs in the absence of oxygen.) At the same time, however, carbon monoxide is removed from the atmosphere by the activities of certain soil microorganisms, so the net result is a harmless average concentration that is less than 0.12–0.15 ppm in the Northern Hemisphere. Because stationary source combustion facilities are under much tighter environmental control than are mobile sources, the principal source of carbon monoxide caused by human activities is motor

Table 9.4. U.S. emission estimates, 1986 (10^{12} g/yr.)

Source	SO_x	NO_x	VOC	CO	Lead	PM
Transportation	0.9	8.5	6.5	42.6	0.0035	1.4
Stationary source fuel	17.2	10.0	2.3	7.2	0.0005	1.8
Industrial processes	3.1	0.6	7.9	4.5	0.0019	2.5
Solid waste disposal	0.0	0.1	0.6	1.7	0.0027	0.3
Miscellaneous	0.0	0.1	2.2	5.0	0.0000	0.8
Total	21.2	19.3	19.5	61.0	0.0066	6.8

Source: USEPA, *National Air Pollutant Emission Estimates 1940–1986, Washington, D.C.*, 1988.

vehicle exhaust, which contributes about 70% of all CO emissions in the United States.

▶ **Key Point:** Soil microorganisms act as a CO sink.

CO also slowly breaks down in the atmosphere to produce carbon dioxide (CO_2). Carbon dioxide and other infrared-absorbing gases in the atmosphere contribute to global warming (the greenhouse effect).

▶ **Key Point:** Methane (CH_4), chlorofluorocarbons (CFCs), water vapor, and nitrous oxide (N_2O) are also greenhouse gases.

Greenhouse gases and global warming. Most serious gardeners understand the operation and importance of a garden greenhouse. The typical garden greenhouse is composed of glass walls and ceilings. These glass partitions are transparent, of course, to short-wave radiation from the sun, which is absorbed by the surfaces and objects inside the greenhouse. Once absorbed, the radiation is transformed into long-wave (infrared) radiation (heat), which is radiated back from the interior of the greenhouse. But the glass does not allow the long-wave radiation to escape; instead, it absorbs the warm rays. With the heat trapped inside, the interior of the greenhouse becomes and remains much warmer than the air outside.

Earth's atmosphere allows much the same greenhouse effect to occur. The short-wave and visible radiation that reaches earth is absorbed by the surface as heat. The long heat waves are then radiated back out toward space, but the atmosphere instead absorbs many of them. This is a natural and balanced process and, indeed, is essential to life as we know it on Earth. The problem comes when changes in the atmosphere radically change the amount of absorption and, therefore, the amount retained. Scientists in recent decades speculate that this may have been happening as various air pollutants cause the atmosphere to absorb more heat. This phenomenon occurs at the local level with air pollution, causing heat islands in and around urban centers.

As mentioned, the main contributors to this effect are the so-called greenhouse gases: water vapor, carbon dioxide, carbon monoxide, methane, volatile organic compounds (VOCs), nitrogen oxides, chlorofluorocarbons (CFCs), and surface ozone. These gases delay the escape of infrared radiation from Earth into space, causing a general climatic warming. Scientists stress that this is a natural process. Indeed, Earth would be 33°C cooler than it is now if the "normal" greenhouse effect did not exist.[8]

The problem with Earth's greenhouse effect is that human activities are now rapidly intensifying this natural phenomenon, perhaps leading to global warming. Debate, confusion, and speculation about this potential consequence are rampant. Scientists are not entirely sure whether the recently perceived worldwide warming trend is due to greenhouse gases or to some other cause or whether it is simply a wider variation in the normal heating and cooling trends they have been studying. If it continues unchecked, however, the process may lead to significant global warming, with profound effects. Human impact on the greenhouse effect is real; it has been detected and measured. The levels of "greenhouse gases" have rapidly increased in recent decades, and this trend is continuing. The rate at which the greenhouse effect is intensifying is now more than five times what it was during the last century.

✔ **Key Point:** Most computer models predict global warming of 1.5°C to 5°C, which would have a profound effect on rainfall, plant growth, and sea levels (rising as much as 1.5 meters).

At the present time, scientists are able to point to six scenarios that could result in long-term global warming and cooling:

1. Changes in Earth's position relative to the sun occur, with higher temperatures when the two are closer together and lower when further apart.
2. Major catastrophes occur (meteor impacts or massive volcanic eruptions) and throw pollutants into the atmosphere, blocking out solar radiation.
3. Changes in albedo (reflectivity of Earth's surface) occur. If Earth's surface were more reflective, for example, the amount of solar radiation radiated back toward space (instead of being absorbed) would increase, lowering temperatures on earth.
4. The amount of radiation emitted by the sun changes.
5. The shape and relationship of the land and oceans change.
6. The composition of the atmosphere changes.

The true effect of increased greenhouse gases is difficult to predict because of:

• The role of the oceans
• The influence of clouds
• Biofeedback mechanisms

Sulfur Dioxide (SO_2)

Sulfur dioxide is a colorless gas possessing the sharp, irritating, pungent odor of burning rubber. On a global basis, nature and human activities produce sulfur dioxide in roughly equivalent amounts. Its natural sources include volcanos, decaying organic matter, and sea spray, while anthropogenic sources include combustion of sulfur-containing coal and petroleum products and smelting of nonferrous ores.

✔ **Key Point:** Background levels of SO_2 are very low, with typical concentrations in the range from 24 to 90 pptv. In remote areas relatively unaffected by pollutant sources, concentrations are typically <5 ppbv.

According to the World Resources Institute and Internal Institute for Environment and Development (WRI & IIED), in industrial areas much more sulfur dioxide comes from human activities than from natural sources.[9] Sulfur-containing substances are often present in fossil fuels; SO_2 is a product of combustion that results from the burning of sulfur-containing materials. The largest single source of sulfur dioxide is from the burning of fossil fuels to generate electricity. Thus, near major industrialized areas, it is often encountered as an air pollutant.

✔ **Key Point:** Approximately 100 million metric tons of sulfur per year enter the global atmosphere by way of human activity, primarily the burning of coal and fuel oil.

In the atmosphere, sulfur dioxide converts to sulfur trioxide (SO_3) and sulfate particles (SO_4). Sulfate particles restrict visibility, and in the presence of water form sulfuric acid (H_2SO_4), a highly corrosive substance that also lowers visibility.

Global output of sulfur dioxide has increased sixfold since 1900. Most industrial nations, however, since 1975–1985, have lowered sulfur dioxide levels by 20% to 60% by shifting away from heavy industry and imposing stricter emission standards. Major sulfur dioxide reductions have come from burning coal with a lower sulfur content and from using less coal to generate electricity.[10]

Two major environmental problems have developed in highly industrialized regions of the world where the atmospheric sulfur dioxide concentration has been relatively high: sulfurous smog and acid rain. Sulfurous smog is the haze that develops in the atmosphere when molecules of sulfuric acid accumulate, growing in size as droplets until they become sufficiently large to serve as light scatterers. The second problem, acid rain, is precipitation contaminated with dissolved acids such as sulfuric acid. Acid rain has posed a threat to the environment by causing certain lakes to become devoid of aquatic life.

✔ **Key Point:** Sulfur dioxide primarily affects the respiratory tract, producing increased mucous secretion, irritation, and increased air resistance. It is also harmful to plants, causing leaf necrosis and/or chlorosis (yellowing) of the green portions of the plant.

Nitrogen Oxides (NO_x)

The nitrogen oxides comprise a group of seven oxides of nitrogen—NO, NO_2, NO_3, N_2O, N_2O_3, N_2O_4, and N_2O_5—but only two are important in the study of air pollution: nitric oxide (NO) and nitrogen dioxide (NO_2).

✔ **Key Point:** Nitrous oxide (N_2O) is generated by microbes and is relatively unreactive, although it could contribute to ozone depletion. Colorless, odorless nitric oxide (NO) and pungent red-brown nitrogen dioxide (NO_2) are collectively know as NO_x and are very significant in polluted air.

Nitric oxide is a colorless, slightly sweet, relatively nontoxic gas, which is produced by both natural and human actions. Soil bacteria are responsible for the production of most of the nitric oxide produced naturally and released to the atmosphere. Within the atmosphere, nitric oxide readily combines with oxygen to form nitrogen dioxide, and, together, those two oxides of nitrogen are usually referred to as NO_x (nitrogen oxides). NO_x is formed naturally by lightning and by decomposing organic matter. Approximately 50% of anthropogenic NO_x is emitted by motor vehicles and about 30% comes from power plants, with the other 20% produced by industrial processes.

Scientists distinguish between two types of NO_x—thermal and fuel—depending on its mode of formation. Thermal NO_x is created when nitrogen and oxygen in the combustion air (such as those within internal combustion engines) are heated to a high enough temperature (above 1,000 K) to cause nitrogen (N_2) and oxygen (O_2) in the air to combine. Fuel NO_x results from the oxidation (i.e., combines with oxygen in the air) of nitrogen contained within a fuel such as coal. Both types of NO_x generate nitric oxide first, and then, when vented and cooled, a portion of nitric oxide is converted to nitrogen dioxide. Although both thermal and fuel NO_x can be significant contributors to the total NO_x emissions, fuel NO_x is usually the dominant source, with approximately 50% coming from power plants (stationary sources) and the other half released by automobiles (mobile sources).

Nitrogen dioxide is about four times more toxic than nitric oxide and is a much more serious air pollutant. Nitrogen dioxide, at high concentrations, is believed to contribute to heart, lung, liver, and kidney damage. In addition, because nitrogen dioxide occurs as a brownish haze (giving smog its reddish-brown color), it reduces visibility. When nitrogen dioxide combines with water vapor in the atmosphere, it forms nitric acid (HNO_3), a corrosive substance that, when precipitated out as acid rain, causes damage to plants and corrosion of metal surfaces.

NO_x levels rose in several countries and then leveled off or declined during the 1970s. During this same time frame, levels of nitrogen oxide have not dropped as dramatically as those of sulfur dioxide, primarily because a large part of total NO_x emissions comes from millions of motor vehicles, while most sulfur dioxide is released by a relatively small number of emission-controlled, coal-burning power plants.

Acid Rain

Most rainfall is slightly acidic because of decomposing organic matter, the movement of the sea, and volcanic eruptions, but the principal factor is atmospheric carbon dioxide, which causes carbonic acid to form. *Acid rain* (pH $<$ 5.6) is a broad term used to describe several ways that acids fall out of the atmosphere. A more precise term is *acid deposition,* which has two parts: wet (acid precipitation) and dry (dry deposition, dry gases). Acid rain is produced by the conversion of the primary pollutants, sulfur dioxide and nitrogen oxides, to sulfuric acid and nitric acid, respectively. These processes are complex and are dependent on the physical dispersion processes and the rates of the chemical conversions.

✔ **Key Point:** Acid rain is a serious environmental problem that affects large parts of the United States and Canada.

Contrary to popular belief, acid rain is not a new phenomenon, nor does it result solely from industrial pollution. Natural processes—volcanic eruptions and forest fires, for example—produce and release acid particles into the air. The burning of forest areas to clear land in Brazil, Africa, and other countries also contributes to acid rain. However, the rise in manufacturing that began with the industrial revolution dwarfs all other contributions to the problem.

The main culprits are emissions of sulfur dioxide, from the burning of fossil fuels such as oil and coal, and nitrogen oxide, formed mostly from internal combustion engine emissions, which is readily transformed into nitrogen dioxide. These mix in the atmosphere to form sulfuric acid and nitric acid.

$$SO_2 + 1/2O_2 + H_2O \text{ several steps} \rightarrow 2H^+ + SO_4^{-2} \text{ (aq)}$$
$$2NO_2 + 1/2O_2 + 2H_2O \text{ several steps} \rightarrow 2\{H^+ + NO_3^-\}\text{(aq)}$$

In abating the affects of atmospheric acid deposition, Earth's ecosystems are not completely defenseless; they can deal with a certain amount of acid through natural alkaline substances in soil or rocks that buffer and neutralize acid. The American Midwest and southern England are areas with alkaline soil (limestone and sandstone), which provides some natural neutralization. Areas with this soil and those laid on granite bedrock, however, have little ability to neutralize acid rain.

Despite intensive research into most aspects of acid rain, scientists still have many areas of uncertainty and disagreement. That is why the progressive, forward-thinking countries emphasize the importance of further research into acid rain. And that is why the Clean Air Act of 1990 was strengthened to initiate a permanent reduction in SO_2 levels.

▶ **Key Point:** Inorganic air pollutants contribute to many environmental problems, including: poor visibility, acid rain, increased respiratory problems, and the greenhouse effect.

CFCs

CFCs (chlorofluorocarbons) are a class of human-made chemicals known by such tradenames as Freon, Generton, and Isotron.

CFCs are highly volatile compounds, contributing to air pollution. CFCs are unusual because they do not break down when vaporized into the atmosphere. Instead, they rise slowly through the atmosphere, taking six to eight years to reach the stratosphere. Here CFCs can reside for more than a hundred years.

CFCs are implicated in two major threats to the global environment: the greenhouse effect and the reduction of the ozone layer. CFCs contribute to the greenhouse effect, warming the atmosphere by trapping heat, which is then radiated back into the atmosphere. CFCs are more than 10,000 times as effective at trapping this radiated heat than is carbon dioxide.

CFCs have also been shown to contribute to the depletion of the protective ozone layer in the atmosphere. Depletion of the ozone layer permits greater amounts of ultraviolet radiation to reach the earth. The increase in ultraviolet radiation affects human health by increasing the likelihood of developing skin cancer and cataracts and may depress the human immune system. Increased ultraviolet radiation reduces crop yields, depletes marine fisheries, damages construction materials, and increases smog.

ORGANIC AIR POLLUTANTS

Organic air pollutants fall into two categories:

- Direct pollutants that cause cancer, etc.
- Secondary pollutants that contribute to photochemical smog

Photochemical Smog

By far the most damaging photochemical air pollutant is ozone. Other photochemical oxidants (peroxyacetyl nitrate [PAN], hydrogen peroxide [H_2O_2], aldehydes, acrolein, peroxybenzoyl nitrate [PBzN], and formaldehyde) play minor roles. All of these are secondary pollutants because they are not emitted but are formed in the atmosphere by photochemical reactions involving sunlight and emitted gases, especially NO_x and hydrocarbons.

✔ **Key Point:** The conditions needed for the formation of the smog are present in all modern cities. They include sunlight, hydrocarbons, nitrogen oxides, and particulates that act as catalysts.

On rare occasions, it is possible for upper stratospheric ozone (good ozone) to enter the lower atmosphere (troposphere). Generally, this phenomenon occurs only during an event of great turbulence in the upper atmosphere. On rare incursions, atmospheric ozone is formed and consumed by endogenous photochemical reactions, which are the result of the interaction of hydrocarbons, oxides of nitrogen, and sunlight, which produces a yellowish-brown haze commonly called smog (i.e., Los Angeles-type smog). In its very simplest terms, we can express the formation of photochemical smog as

$$\text{Hydrocarbons } NO_x + \text{sunlight} \rightarrow \text{photochemical smog}$$

✔ **Interesting Point:** On 9 December 1952, foggy conditions developed over London. It was very cold, so most houses kept fires burning, with coal as the major fuel. The smoke from these fires mixed with the fog and was unable to disperse, resulting in smog that persisted for four days. The pH of air during the Great London Smog was as low as 1.6. During this period some four thousand more people died than expected at this time of the year. Most of these additional deaths were due to respiratory disorders.

Although the incursion of stratospheric ozone into the troposphere can cause smog formation, the actual formation of heavy smog involves a complex group of photochemical interactions. These interactions are between anthropogenically emitted pollutants (NO and hydrocarbons) and secondarily produced chemicals (PAN, aldehydes, NO_2, and ozone). Note that the concentrations of these chemicals exhibit a pronounced diurnal pattern, depending on their rate of emission and on the intensity of solar radiation and atmospheric stability at different times of the day.

> ✔ **Key Point:** Photochemical smog has a long history, dating back at least to the 1860s in Los Angeles. Its characteristics are low visibility, eye irritation, and deterioration of materials. By definition, smog is identified by visibility below three miles and moderate to severe eye irritation when the relative humidity is below 60%. Photochemical smog requires UV light, hydrocarbons, and NO_x.

The time line for the presence of various air pollutants in the atmosphere of Los Angeles shows that in bright sunlight, the NO is photochemically oxidized to NO_2, resulting in a decrease in NO concentration and a peak of NO_2 between 7:00 and 9:00 a.m. Photochemical reactions involving NO_2 produce O atoms, which react with O_2 to form O_3. These result in a net decrease in NO_2 concentration and an increase in O_3 concentration, peaking between noon and 3:00 p.m. Aldehydes, also formed photochemically, peak earlier than O_3. As the day proceeds, the various gases decrease in concentration as they are diluted by fresh air masses or are consumed by photochemical reactions. This cycle is typical of an area that experiences photochemical smog and is repeated daily.[11]

The considerable range of the estimates for the Northern Hemisphere reflects uncertainty in the calculation of the ozone fluxes. On average, stratospheric incursions account for about 18% for the total ozone influx to the troposphere, while endogenous photochemical production accounts for the remaining 82%. About 31% of the tropospheric ozone is consumed by oxidative reactions in vegetative and inorganic suffocates at ground level, while the other 69% is consumed by photochemical reactions in the atmosphere.[12]

Most organic pollutants in the atmosphere come from natural sources. For example, 85.7% of methane is naturally generated via anaerobic bacteria and domestic animals. The other 14.3% is the result of human activity.

Methane is a major source of O^3 and CO in the troposphere and water vapor in the stratosphere.

$$CH_4 + HO^* \rightarrow H_3C^* + H_2O$$

Other natural sources of atmospheric hydrocarbons are plants, which release ethylene and a variety of terpenes and esters.

Human-made hydrocarbon sources, because of the burning of fossil fuels, are the number one class of organic air pollutants. These common hydrocarbon air pollutants are produced in quantities exceeding 1,000,000,000 kg/yr.

Aromatic hydrocarbons are widely used in industry and are key components of unleaded gasoline.

Hydrocarbons are typically oxidized (by $h\nu$) to *aldehydes* and *ketones*. Billions of kilograms of other important *carbonyl* compounds are produced annually. Aldehydes are second only to NO_2 as a source of photochemically produced free radicals.

Many *alcohols* also rank among the top fifty manufactured chemicals, the most volatile of which have been identified as atmospheric pollutants (e.g., methanol, and ethanol).

✔ **Key Points:** Alcohols with high water solubility and/or low volatility are quickly "scavenged" from the atmosphere.

Phenol (an aromatic alcohol) is also in the "top 50," and is another by-product of burning coal and making coke. Other common aromatic alcohols have also been identified as atmospheric pollutants.

The three most common *organic halide pollutants* of the atmosphere are methyl chloride, methyl chloroform, and carbon tetrachloride. The destruction of the ozone by CFCs and halides has already been discussed. Significant *organonitrogen pollutants* include:

• Amines (odor!)
• Dimethylformamide
• Acrylonitrile (top 50)

Cigarette smoke, burning vegetation, and coke ovens release heterocyclic nitrogen compounds.

Peroxyacetylnitrate (PAN) and *peroxybenzoylnitrate* (PBN) are serious organic pollutants produced by the photochemical oxidation of hydrocarbons.

Polyaromatic hydrocarbons (PAHs) are the most notorious organic particulates in our atmosphere (up to ~ 20 ug/m^3). They are produced by the incomplete combustion of fossil fuels and cigarettes (− 100 u/g/m^3). These compounds are typically found sorbed to soot (another PAH).

Other minor but often troublesome organic pollutants include:

• Ethers (fossil fuels, THF)
• Oxides (ethylene and propylene oxide)
• Carboxylic acids (photochemical oxidation)
• Organosulfur compounds (odor!)

Samples of smog have been shown to contain PAN, aldehydes, ketones, alkyl nitrates, and alkyl nitrites, primarily from the photoxidation of hydrocarbons. These air samples also contained inorganic pollutants such as ozone and nitric acid.

✔ **Key Point:** *Smog* = hydrocarbons + *hv* + NO$_x$

✔ **The bottom line on organic air pollutants:** They contribute to many health and environment problems, including smog, respiratory difficulties, poor visibility, damage to materials, toxicity to plants, and increased cancer risks.

Chemistry and the Clean Air Act (CAA)

Most air quality regulations are recent. For example, in the United States, the first attempt at regulating air quality came about through passage of the Air Pollution Control Act of 1955. This act was a step forward but that's about all; it did little more than move the country *toward* effective legislation. Revised in 1960 and again in 1962, the act was supplanted by the Clean Air Act (CAA) of 1963. CAA 1963 encouraged state, local, and regional programs for air pollution control but reserved the right of federal intervention should pollution from one state endanger the health and welfare of citizens residing in another state. In addition, CAA 1963 initiated the development of air quality criteria upon which the air quality and emissions standards of the 1970s were based.

The Clean Air Act Amendments signed by George Bush in November of 1990 were the first clean air legislation in twenty years. With these acts on their way to being enforced, all Americans may have the same basic health and environmental protections. The acts allow the individual states to have stronger controls on air pollution but are not allowed to have a weaker regulations than those set for the rest of the country. Specifically, the new law:

- Encourages the use of market-based principles and other innovative approaches such as performance-based standards and emission banking and trading
- Promotes the use of clean low-sulfur coal and natural gas, as well as the use of innovative technologies to clean high-sulfur coal through the acid rain program
- Reduces enough energy waste and creates enough of a market for clean fuels derived from grain and natural gas to cut dependency on oil imports by a million barrels a day
- Promotes energy conservation through an acid rain program that gives utilities flexibility to obtain needed emission reductions through programs that encourage customers to conserve energy

Additional Reading

Hobbs, P.V. *Introduction to Atmospheric Chemistry.* Cambridge University Press, 2000.
Seinfeld, J. H., and S. N. Pandis. *Atmospheric Chemistry and Physics: From Air Pollution & Climate Change.* New York: Wiley-Interscience, 1997.
Wallace, J. M., and P. V. Hobbs. *Atmospheric Science: An Introductory Survey.* Academic Press, 1997.

Summary

- Normal air consists of nitrogen (79%), oxygen (20%), and dust (solid), water (solid, liquid, and gas), carbon dioxide, and trace elements (1%).
- The lowest layer of the atmosphere is the troposphere.

- Acid rains are normal, but human pollution has increased the acid concentration of rain to more than ten times that of natural rains.
- Carbon dioxide seems to be a harmless pollutant in our air.
- Carbon monoxide is a human-made pollutant from automobiles. It is a poisonous gas that can cause death.
- The automobile is a major air polluter. It releases carbon monoxide, oxides of nitrogen, lead, and cancer-causing chemicals.
- The major categories of atmospheric chemical species are inorganic oxides, oxidants, reductants, organics, photochemically active species, acids, bases, salts, and unstable reactive species.
- Gaseous pollutants that enter the atmosphere in the greatest quantities are CO, SO_2, NO, and NO_2.
- Three relatively reactive and unstable species encountered in the atmosphere that are strongly involved with atmospheric chemical processes are electronically excited molecules, free radicals, and ions.
- The photochemical dissociation of nitrogen dioxide, NO_2, can produce reactive O atoms that can react with oxidizable molecules.
- A photochemical atmosphere polluted by nitrogen oxides and hydrocarbons generates strong oxidant molecules.
- The most prominent inorganic oxidant in the atmosphere is ozone.
- The two reactions by which stratospheric ozone is produced are $O_2 + hv \rightarrow O + O$ and $O + O_2 \, M \rightarrow O_3 + M$.
- The effects of pollutants in the atmosphere may be divided between direct effects and the formation of secondary pollutants, such as photochemical smog.
- The fact that most organics (hydrocarbons) in the atmosphere come from natural sources is primarily the result of the release of huge quantities of methane.
- The fluorine-containing air pollutants with the greatest potential for damage to the atmosphere are the CFCs.

New Word Review

Aerosols—microscopic solids and liquids suspended in the air.

Acid rain—rainfall with a greater acidity than normal rain. Normal rain has a pH of about 5.6. Acid rain results from air contaminants of sulfur dioxide and nitrogen oxides that dissolve in water to form acids.

Air—the mixture of gases that fills the lower portion of the atmosphere. The dry air consists of (by volume) 78% nitrogen, 21% oxygen, 1% argon, and trace amounts of carbon dioxide, helium, hydrogen, methane, etc. On the average, water vapor to 0.7% is also present.

Atmosphere—the envelope of gases surrounding the earth. It extends up to 600 miles over the entire earth surface.

Chlorofluorocarbons (CFCs)—a class of hydrocarbon derivatives in which chlorine and fluorine are substituted for some or all of the hydrogens. They are widely used in consumer products. They are implicated in ozone depletion and the greenhouse effect.

Clean Air Act (CAA)—the basic federal air pollution control statute. It was first passed in 1963 and has been amended periodically. It provides for the national ambient air quality standards, the state implementation plant process, the prevention of significant deterioration program, and national emission standards for hazardous pollutants. The 1990 amendments included provision pollutants of global concerns and stricter auto emission standards.

Dispersion—the scattering process similar to diffusion. In air pollution, dispersion indicates the combined action of advection and diffusion.

Emission—the release of pollutants in atmosphere.

Greenhouse effect—the trapping of sunlight as heat. The sunrays pass through atmosphere and are absorbed by the earth. The earth radiates this energy as heat (infrared) waves. The carbon dioxide in the air, along with some other gases, absorbs most of these heat waves and reradiates them towards the earth and to space. The atmosphere, thus, acts like a greenhouse, hence, the name. Higher concentration of carbon dioxide will reradiate more heat waves, increasing the earth temperature. This will have many environmental consequences.

Pollution—the contamination of the environment by the introduction of a poisonous substance, or by changing the concentration of an existing substance.

Nitrogen oxides—NO, NO_2, NO_3, N_2O, N_2O_3, N_2O_4, and N2O5. The first two, nitric oxide and nitrogen dioxide, are the primary air pollutants.

Ozone depletion—the destruction of ozone molecules in the ozone layer by chemical reactions with compounds released by humans, such as CFCs and halons.

Particulate—fine solid particles or liquid droplets (not gases) such as smoke, mist, fumes, smog, and aerosols, found in dirty air stream.

Peroxyacetyl nitrate (PAN)—a component of photochemical smog created by reaction involving sunlight, hydrocarbons, and oxides of nitrogen. It causes eye irritation and vegetation injuries.

Photochemical oxidants—the components of photochemical smog. These include ozone, nitrogen dioxide, PAN, and oxygenated hydrocarbons. The highest concentration is of ozone. EPA has set the standards for photochemical oxidants.

Photochemical smog—a light haze formation consisting of photochemical oxidants caused by the effect of the sunlight on automobile exhaust gases. It causes breathing problems, coughing, chest soreness, and eye irritation. It damages plants and cracks rubber. Thermal inversion leads to formation of smog close to the surface.

Pollutant—any chemical substance or physical agent (heat, sound, electromagnetic radiation) introduced to the environment in an amount that threatens human health, wildlife, plants, or the orderly functioning and/or human enjoyment of any aspect of the environment.

Primary air pollutant—a pollutant in the ambient air that is hazardous in the same form in which it is released into the atmosphere; for example, carbon monoxide. The air pollutant is Secondary Air Pollutant, made up of substances formed in the atmosphere by reactions of primary air pollutants; for example, ozone.

Stratosphere—the atmosphere surrounding the earth is stratified in zones with distinct temperature variations. The second layer above the troposphere to about 30 mi is the stratosphere. Here temperature increases with altitude. The protective ozone layer is located in the stratosphere.

Ultraviolet (UV) Radiation—the portion of the electromagnetic spectrum that extends from the violet band of visible light (wavelength of about 0.4 micrometer) to the X-rays (wavelength of about 0.001 micrometer). Radiation in this range does not ionize matter and hence is termed *nonionizing*. The spectrum between wavelengths 0.001 and 0.16 by micrometer does not transmit through air and is of little significance to the environment.

Chapter Review Questions

9.1. Is the atmosphere limitless in its ability to absorb all airborne pollutants without affecting everyone and everything that breathes?

9.2. _____ pollutants are those entering the air directly in harmful form.

9.3. _____ pollutants are those created by physical or chemical changes after they enter the atmosphere.

9.4. _____ directly affects plants and animals and can be further oxidized in air to form sulfur trioxide.

9.5. When they escape into the atmosphere, _____ react with the ozone layer in the stratosphere in such a way that ozone is destroyed.

9.6. _____ or brown smog results from reactions between pollutants and sunlight.

9.7. The _____ established primary standards to protect human health.

9.8. _____ comes from sulfur and nitrogen oxides emitted from burning fossil fuels, primarily from power plants.

9.9. Contamination of the environment: _____

9.10. The main components of our air are _____ and _____.

9.11. _____ are microscopic liquid and solid particles suspended in the air.

9.12. A cloud is formed when droplets of water condense on _____.

9.13. Human-made pollution takes place when _____ is in the air.

9.14. _____ is a metal pollutant released into the air by older cars.

9.15. The trapping of sunlight as heat: _____

9.16. Microscopic solids and liquids suspended in the air: _____

9.17. Pollution of the environment is a(n) _____ development.

9.18. The worst offender in air pollution is the: _____.

9.19. Acid rain is _____.

9.20. Carbon dioxide can produce a _____ and warm the earth.

Notes

1. T. E. Graedel and P. J. Crutzen, *Atmosphere, Climate, and Change* (New York: Scientific American, 1995).

2. S. E. Manahan, *Environmental Science and Technology* (Boca Raton, Fla.: CRC Press, 1997).

3. Manahan, *Environmental Science and Technology*, 267.

4. From *Air Pollutants*, http://www.doc.mmu.ac.uk/aric/eae/Air_Quality/Older/Air_Pol lutants.html.

5. D. R. Hastie, *Atmospheric Particulate Matter.* http://www.cac.yorku.ca/people/hastie/ (accessed December 22, 2002).

6. G. M. Masters, *Introduction to Environmental Engineering and Science* (Englewood Cliffs, N.J.: Prentice-Hall, 1991), 292.

7. V. J. Schaefer and J. A. Day, *Atmosphere: Clouds, Rain, Snow, Storms* (Boston: Houghton Mifflin, 1981).

8. J. E. Hansen, et al., "Climate Sensitivity to Increasing Greenhouse Gases," in *Greenhouse Effect and Sea Level Rise: A Challenge for This Generation,* ed. M.C. Barth and J. G. Titus (New York: Van Nostrand Reinhold, 1986).

9. WRI & IIED, *World Resources 1988–1989* (New York: Basic Books, 1988).

10. J. J. MacKenzie and T. El-Ashry, *Ill Winds: Airborne Pollutants' Toll on Trees and Crops* (Washington, D.C.: World Resource Institute, 1988).

11. P. Urone, "The Primary Air Pollutants—Gaseous: Their Occurrence, Sources, and Effects," in *Air Pollution*, vol. 1, ed. A. C. Stern (New York: Academic Press, 1976).

12. B. Freedman, *Environmental Ecology* (New York: Academic Press, 1989).

CHAPTER 10

Soil Chemistry

When we change oil in our cars and dump the dirty oil in our backyard soil, is this a problem of global proportions? The simple answer is yes. The compound answer is that the immediate human impacts on the land reach a global magnitude, if they do, in a patchwork and cumulative fashion. They are "changes that are local in domain, but which are widely replicated and which in sum constitute change in the whole human environment. Individually, they pose issues of less than global domain . . . but they add up to a global effect."[1]

Topics in This Chapter

- The Soil Problem
- Physical and Chemical Properties of Soil
- Subsurface Fate and Transport
- Origins of Soil Pollutants
- Soil Pollution Remediation
- CERCLA/RCRA

Environmental and occupational health and safety professionals work in broad-spectrum fields. Preservation of safe, healthy lives of workers and the public requires more than the careful management of the workplace; it also requires careful management of all the primary natural resources: air, water, and soil. Each field of scientific endeavor has the capacity to contribute understanding to one or more aspects of these resources. As each of us focuses on our individual niche of the universe, we become increasingly aware of the interrelationships between all the various natural (and human-made) environmental parameters. Constant interplay occurs between the atmosphere, surface water, groundwater, and soil. No one portion of our earthly life support system is independent. Plants and animals depend on these complex support systems. Soils, as Winegardner points out, "are the bridge between mineral matter and life."[2]

The Soil Problem

Soil contamination is a major environmental concern not only throughout the United States but worldwide. "The impact on soil from industrial and agricultural practices, management of Superfund sites, exploration and production, and mining and nuclear industrial practices, among others, remains difficult to assess. Certainly petroleum-contaminated soil affects the largest number of sites and the largest total volume of impacted material. The overall amount of contaminated soil generated can be staggering. For example, in some states such as Oklahoma, contaminated soil accounts for about 98% of the waste generated as a one-time occurrence."[3]

As our concern for the environment increases, it is comforting to recognize that soil, when properly used, can offer an unlimited potential for disposal and recycling of waste materials. Knowledge of physical and chemical reactions is more important to environmental practitioners today than ever before.

Regardless of the origin, most soils consist of four basic components: mineral matter, water, air, and organic matter. As mentioned, soils are the bridge between mineral matter and life.

In order to gain understanding of the "soil problem" currently confronting us, it is necessary to briefly describe a few environmental problem areas that contribute to the soil problem. These problem areas include underground storage tanks (USTs), chemical sites, oil field sites, geothermal sites, manufactured-gas plants, mining sites, and environmental terrorism.

UNDERGROUND STORAGE TANKS (USTS)

Petroleum contamination is commonly associated with USTs. Recent estimates have ranged from five to six million, but no one is quite sure just how many USTs containing petroleum products or hazardous materials are in use in the United States. Compounding the issue, no one can guess how many USTs are no longer being used and have been abandoned. Abandoned tanks often still hold (or held) some portions of their contents, which may have been oozing out, fouling water, land, and air. Another problem is that older USTs that are not leaking today will probably leak eventually. One thing is certain; environmental contamination from leaking USTs poses a significant threat to human health and the environment.

Besides the obvious problem of fouling the environment, many of these leaking USTs also pose serious fire and explosion hazards. The irony is that USTs came into common use primarily as a fire and explosion prevention measure (presumably, the hazard was buried under the ground—out of sight, out of mind, out of harm's way). Today, however, the hazards we "buried" to protect ourselves from are finding ways to present hazards in a different manner.

The problem with leaking USTs goes beyond fouling the environment (especially groundwater, which 51% of the U.S. population relies on for drinking water) and presenting fire and explosion hazards. Products released from these leaking tanks can damage sewer lines and buried cables and poison our crops.

CHEMICAL SITES

A 1979 survey (called the Eckhardt Survey after the legislator who initiated it) of 53 of the largest chemical manufacturing companies in the United States, conducted for the period 1950 to 1979, reported almost 17 million tons of organically generated waste disposed of. Of this total, a little over 10 million tons were untreated (i.e., ponds, landfills, lagoons, and injection wells). A little under 0.5 million tons were incinerated, and a little over 0.5 were either recycled or reused. Not addressed is the volume of contaminated soil generated as a one-time occurrence, as is typical of any remediation activity.[4]

OIL FIELD SITES

Another source of rather large volumes of hydrocarbon-contaminated soil is related to production in past and existing oil fields. Experience has shown that it is relatively easy to describe the fate of petroleum in soils in merely qualitative terms. For example, it is clear that much of volatile petroleum products such as gasoline are lost by evaporation; normal alkanes are subject to fairly rapid biodegradation; aromatic hydrocarbons, particularly those of lower molecular weight, are very susceptible to dissolution into water and may thus cause contamination of water supplies in the locality. It is, however, much more difficult to make rigorous, scientifically justifiable statements in which these processes are described in quantitative terms. In very few cases has it been possible to state, for example, that in a period of one year a certain percentage of a particular mass of oil spill was lost by evaporation, a certain percentage by dissolution, and another percent by biodegradation, and the rest was altered by photolysis. One thing is certain, however; oil field sites, both historically and in the present, contribute to the overall volume of contaminated soil generated.[5]

GEOTHERMAL SITES

The U.S. Department of Energy defines the capture of *geothermal energy* as "when the heat contained within the earth is recovered and put to useful work." Geothermal energy is a very efficient resource for heating and cooling buildings, drying agricultural products, and processing heating for industry.[6] High-temperature (above 150°C) geothermal resources are used in electric power generation. However, one of the biggest problems in using geothermal energy is extracting it. To obtain geothermal power efficiently, heat energy must be concentrated in a small area. This may be done by underground reservoirs of hot water or steam that can be funneled into a drill hole.

Geothermal energy is used best in the generation of electricity. In geothermal power plants, hot water is turned into steam, which is then used to power a turbine. The mechanical energy from the turbine is then converted to electricity by a generator. Another source of geothermal energy is "hot and dry," from subsurface rocks.[7]

▶ **Key Point:** The U.S. Department of Energy has a Geothermal Energy Technical Site that goes into great detail on the growth of the geothermal industry. It can be located at http://www.eren.doe.gov/geothermal/.

Proponents of geothermal use point out that geothermal energy is clean, reliable, readily available, and easy on the land—a very beneficial form of energy with a strong future ahead of it.

The use of geothermal energy is a double-edged sword, however. Although geothermal sources eliminate the air pollution associated with combustive electrical generation, they are not without problems. Gases in the steam and dissolved minerals in the hot water cause equipment to erode rapidly. The cooling and disposal of the water—mineral-rich brine—from the spring may also present a problem. Moreover, there is evidence that the groundwater in areas where geothermal energy is used must be replaced to prevent subsidence.

✔ **Key Point:** Geothermal-generated brine is a mineralized fluid composed of warm to hot saline waters containing calcium, sodium, potassium, chloride, and minor amounts of other elements.

MANUFACTURED-GAS PLANTS (MGPS)

Manufactured-gas plants (MGPs) have operated since the late 1890s, with many industrial sites having undergone redevelopment.

✔ **Key Point:** Recent surveys estimate that about three thousand MGP sites exist throughout the U.S.

Manufactured-gas plants produce a variety of largely hazardous waste products, almost all of which are found in what is referred to today as coal tar (and associated waste products). Among the toxic substances found in coal tar are:

- Aromatic hydrocarbons—consisting mainly of phenols and cresols
- Monocyclical aromatic hydrocarbons (MAHs)—the so-called BTEX series: *b*enzene, *t*oluene, *e*thylbenzene, and *x*ylene
- Duocyclical aromatic hydrocarbons (DAHs)—consisting mainly of naphthalene and the light oils
- Polycyclical aromatic hydrocarbons (PAHs)—the coal tars and medium and heavy oils
- Others—the concentrated forms of trace minerals found in the coal, including cyanides, sulfur, and some heavy metals (arsenic, chromium, lead, etc.)

MINING SITES

Mining operations often create land and water pollution problems. Erosion-causing sediment pollution is surface mining's most obvious and most thoroughly documented problem.

The effect that mining sediments and mining wastes (from mining, milling, smelting, and leftovers) had and are having on soil is less known because no serious study of the problem has occurred. Typical mining wastes include acid from oxidation of naturally occurring sulfides; asbestos from asbestos mining and milling operations; cyanide from precious-metal heap-leaching operations; leach liquors from copper-dump leaching operations; metals from mining and milling operations; and radionuclides (e.g., radium) from uranium and phosphate mining operations.

Acid mine drainage as a soil contaminant source is well known and well documented. When oxygen and water react with sulfur-bearing minerals to form sulfuric acid and iron compounds, acid formations occur. These compounds may directly affect plant life that absorbs them. Acid mine drainage can also indirectly affect the flora of a region by changing the nature of the soil minerals and microorganisms in the area.

The solid waste by-products of mining cause other problems. Metals are always mixed with other materials in mining. These materials usually have little commercial value and disposal becomes problematic. The unsightly piles of rock and rubble are prone to erosion. Leaching releases waste materials that allow environmental poisons to enter the soil.

ENVIRONMENTAL TERRORISM

Before 1991 (the Gulf War) and 9/11, the general public barely acknowledged the potential for environmental terrorism on a regional scale. Thereafter, it could not be ignored. The deliberate destruction of oil wells and refineries or chemical sites, alone, resulted in soil pollution on a catastrophic scale in the Gulf War.

Physical and Chemical Properties of Soil

From the environmental professional's point of view (regarding land conservation and methodologies for contaminated soil remediation through reuse and recycling), 10 major chemical/physical properties of soil are of interest:

- Cation-exchange capacity (CEC)
- pH
- Salinity
- Color
- Texture
- Tilth
- Water-holding capacity
- Drainage
- Depth
- Slope

CATION-EXCHANGE CAPACITY (CEC)

Cation-exchange capacity (CEC) is the soil's capacity to hold cations. More specifically, CEC is defined as the sum of positive ($+$) charges of the adsorbed cations that a soil can adsorb at a specific pH. Soil particles are composed of silicate and aluminosilicate clay. These particles are negatively charged colloids. Cations are bound ionically to the surface of these colloid particles. A cation is a positively charged ion; for example H^+, Ca^{++}, Mg^{++}, K^+, NH_4^+, and Na^+ are all cations.

✔ **Key Point:** CEC is expressed as meg/100 g of soil.

Typical CEC values by soil type are listed in table 10.1.

✔ **Key Point:** CEC increases as the clay content and the organic matter increases in a soil.

Cations with higher charge densities, i.e., smaller cations, will replace larger cations. For example, H^+ will displace Ca^{++}; Ca^{++} will displace Mg^{++}.

When an ion in the water phase is attracted to a soil surface, it must displace another (already present) cation. For example:

$$(Soil)Mg + Cu^{2+} \rightarrow (Soil)Cu + Mg^{2+}$$

As organic matter decomposes in a soil, protons are produced. The bacteria in the soil causing the decomposition also respire protons (H^+). As the proton content goes up, the protons displace the other bound ions on the surface of the soil particles, and these cations become available for absorption by the plant roots.

✔ **Key Point:** All cations are not created equal!

pH

Recall that pH is the negative logarithm of the hydrogen ion activity. The pH of soil is one of the most important properties involved in plant growth.

All living organisms are sensitive to pH. The plant roots will not function opti-

Table 10.1. CEC values by soil type.

Soil Type	meq/100g soil
Sand	2–4
Loam	7–16
Clay	4–60
Organic	50–300

mally in soils outside a specific pH range unique to that organism. If the pH of the soil is extreme, either alkaline or acid, the plant will die. Soil microorganisms, insects, and other animals present in the rhizosphere (i.e., the zone immediately adjacent to plant roots in which the kinds, numbers, or activities of microorganism differ from that of the bulk soil) are equally sensitive to pH.

✔ **Key Point:** Alkaline soils have pH 7.5–8.5. Acidic soils have pH 4–6.5. Soils with pH values outside these ranges are usually toxic to most plants. Alkaline components are more readily leached from soils.

Soil pH can be adjusted by amendments. Increasing organic matter will decrease pH (increase acidity). Lime can be added to increase pH (increase alkalinity). Certain fertilizers are delivered as acidic or basic solutions. These will also alter soil pH. However, soils have buffering capacity. This means that within their normal range of pH values, they can absorb lots of protons or lots of hydroxyl ions before the pH of the soil water changes. But once the buffering capacity of the soil is reached, the pH of the soil water will change rapidly to toxic extremes. It will also take a lot of new buffering activity to repair the soil to its original pH.

SALINITY

Saline soils are soils with large amounts of soluble salts. Sodic soils are nonsaline soils containing sufficient exchangeable sodium to adversely affect crop production and soil structure under most conditions of soil and plant type. Salts accumulate naturally in some surface soils of arid and semiarid regions because there is insufficient rainfall to flush them from the upper soil layers. The salts are primarily chlorides and sulfates of calcium, magnesium, sodium, and potassium. They may be formed during the weathering of rocks and minerals or brought to the soils through rainfall and irrigation. The total salt concentration of soil, or salinity, is expressed in terms of the electrical conductivity of the water, and it is easily and precisely determined.

COLOR

Just about anyone who has looked at soil has probably noticed that soil color often differs from one location to another. Soil colors range from very bright to dull greasy to a wide range of reds, browns, blacks, whites, yellows, and even greens. Soil color is dependent primarily on the quantity of humus and the chemical form of iron oxides present.

Soil scientists use a set of standardized color charts (the *Munsell Color Book*) to describe soil colors. They consider three properties of color—hue, value, and chroma—in combination to come up with a large number of color chips to which soil scientists can compare the color of the soil being investigated.

For the environmental professional, the color of a soil can give clues to its health, origin, and long-term changes. It can also indicate the color of the parent material. Subsoil color can be a valuable indicator of how well the soil drains, which can be influenced by the topography.

Dark colors in the topsoil usually indicate that the soil has a high organic matter content. We can often see this in prairie grassland soils of the Great Plains in the United States.

The more humus, the blacker the soil. It could mean that the parent material from which soil developed was also black.

✔ **Key Point:** Distinct red and yellow colors usually indicate older, more weathered soils.

TEXTURE

Soil texture, the relative proportions of the various soil separates in a soil, is given and cannot be easily or practically changed significantly. It is determined by the size of the rock particles (sand, silt, and clay particles) or the soil separates (the size groups of mineral particles less than 2 mm in diameter, or the size groups that are smaller than gravel) within the soil. The largest soil particles are gravel, which consist of fragments larger than 2.0 mm in diameter.

Particles between 0.05 and 2.0 mm are classified as sand. Silt particles range from 0.002 to 0.05 mm in diameter, and the smallest particles (clay particles) are less than 0.002 mm in diameter. Though clays are composed of the smallest particles, those particles have stronger bonds than silt or sand, and once broken apart, they erode more readily. Particle size has a direct impact on erodibility. Rarely does a soil consist of only one single size of particle; most are a mixture of various sizes.

As mentioned, soil texture refers to the relative proportions of sand, silt, and clay in a soil. Note that humus content technically has nothing to do with texture. Soils get their textural names—twelve in all—and some specific physical properties from the proportions of these three particle sizes. Eight of them are:

- Loamy Sand—85% sand, 10% silt, and 5% clay.
- Sandy Loam—72% sand, 15% silt, and 13% clay.
- Loam—46% sand, 36% silt, and 18% clay.
- Silt Loam—25% sand, 60% silt, and 15% clay.
- Sandy Clay Loam—65% sand, 9% silt, and 26% clay.
- Clay Loam—36% sand, 32% silt, and 32% clay.
- Sandy Clay—55% sand, 5% silt, 40% clay.
- Clay—17% sand, 17%, and 66% clay.

The other four are sand, silt, silty clay loam, and silty clay.

✔ **Key Point:** A soil's texture has a big influence on its productivity and management needs because it affects tilth, water-holding capacity, drainage, erosion potential, and soil fertility. As explained already, texture usually varies with depth, and the subsoil is usually more clayey than the topsoil.

TILTH (STRUCTURE)

Soil tilth (structure) should not be confused with soil texture—they are different. In fact, in the field, the properties determined by soil texture may be considerably modified by soil tilth. *Soil tilth* refers to the combination or arrangement of primary soil particles into secondary particles (units or peds). Simply stated, soil tilth refers to the way various soil particles clump together. The size, shape, and arrangement of clusters of soil particles called *aggregates* determine the formation of larger clumps called *peds*. Sand particles do not clump because sandy soils lack structure. Clay soils tend to stick together in large clumps. Good soil develops small, friable (easily crumbled) clumps. Soil develops a unique, fairly stable structure in undisturbed landscapes, but agricultural practices break down the aggregates and peds, lessening erosion resistance.

✔ **Key Point:** A soil in good tilth is easily worked, is crumbly, and readily takes in water when dry.

The presence of decomposed or decomposing remains of plants and animals (organic matter) in soil helps not only fertility but also soil structure—especially the soil's ability to store water. Live organisms such as protozoa, nematodes, earthworms, insects, fungi, and bacteria are typical inhabitants of soil. These organisms work to either control the population of organisms in the soil or to aid in the recycling of dead organic matter. All soil organisms, in one way or another, work to release nutrients from the organic matter, changing complex organic materials into products that can be used by plants.

✔ **Key Point:** Factors that influence a soil's tilth include texture, organic matter, and moisture content. Therefore, tilth can vary markedly with changes in moisture content, the amount of humus present, and compaction, especially if it is clayey. Tilth can be improved, which is why farmers plow their fields to break up the clods and add manure to their fields when they can.

WATER-HOLDING CAPACITY

About half of soil's volume is pore space that is occupied by varying amounts of air and water, depending on how wet the soil is. Water is held in the pore spaces in the form of films adhering to the soil particles. Small pores are called *micropores*, while large pores are called *macropores*.

Macropores do not hold water well because the water films become too thick to adhere well to the surrounding soil particles. This water is lost downward as it drains below the root zone by gravity. So macropores allow a soil to retain enough air for root growth, as long as drainage is not impeded.

The films of micropore water, however, resist being drained away by gravity and are responsible for the water-holding capacity of soils. This water is what the roots can tap into and extract for plant use.

✔ **Key Point:** As you might guess, sands, with their large grain size, have a lot of macropores but few micropores. Thus, their water-holding capacity is low, although their drainage is good. Heavy clay soils that have a lot of micropores may have a higher water-holding capacity, but because they have fewer macropores, their drainage is poor. Plants really like a soil that lies in between these two extremes so that their roots have both the air and water that they need.

DRAINAGE

Drainage refers to a soil's ability to get rid of excess water, or water in the macropores, through downward movement by gravity. Topography, texture, tilth, depth, and the presence of compacted layers in the subsoil affect it. With few exceptions (one would be a crop like rice), most plants need fairly good drainage. Without good drainage, plant roots would lack oxygen, nitrogen would be lost, and certain elements like iron and manganese could become soluble enough to injure plant roots. Although more likely in clayey soils, drainage problems also occur on sandy soils where the water table is close to the surface. (The *water table* is the upper surface of the groundwater below which the soil is completely saturated with water.)

✔ **Key Point:** Because soil color can be affected by drainage, it can tell you if your soil is having drainage problems. For example, red, reddish-brown, or yellow subsoil colors generally indicate good drainage. The presence of sufficient air allows the soil's iron and manganese to oxidize or "rust" and give these brighter colors. On the other hand, dull grays and blues indicate a reduced state, with little oxygen, which means poor drainage.

DEPTH

Soil depth refers to how deep, top to bottom, the topsoil plus the subsoil is. Depth can be easily determined by digging a hole. Soils are classified as being deep or shallow, as shown in table 10.2.

✔ **Key Point:** Soil depth is important for plants because deeper rooting means more soil to explore for nutrients and water. Greater soil depth can also mean better drainage, as long as there are no restrictive layers in the subsoil.

Table 10.2. Soil depth classification.

Soil Depth	Measurement (topsoil + subsoil)
Deep	3′
Moderately deep	20″–3′
Shallow	10″–20″
Very shallow	<10″

SLOPE

The slope (the steepness of the soil layer) has a marked influence on the amount of water runoff and soil erosion caused by flowing water. Slope is usually measured in terms of percentages. A 10% slope has 10 feet of vertical drop per 100 feet of horizontal distance. Soil conservation measures become necessary on land with as little slope as 1% to 2% to avoid erosion problems.

Subsurface Fate and Transport

Environmental practitioners not only need to know the basics of soil—that is, soil's chemical and physical properties, they also need to know how soil reacts during chemical spill conditions. More specifically, it is important to understand soil's transport mechanisms that affect the flow regime of contaminants or pollutants.

When a chemical spill or leak occurs, or when pollutants are intentionally discharged or accidentally released in soil, environmental practitioners need to ask what happens to the pollutants. The answer is rather complex and is based on the interactions among and between the chemicals, rock, soil water, and the soil.

The following identifies and explains the properties of the pollutants that are important to the migration, retardation, and transformation—and the ultimate disposition, or fate—of pollutants in soil.

Vapor pressure or volatility. Vapor pressure is the property that determines how easily a pollutant evaporates; the lower the vapor pressure, the more easily a pollutant will evaporate. Volatility is the tendency of a solid or liquid pollutant to pass into the vapor state; how quickly and easily a liquid or solid pollutant evaporates at ordinary temperatures when exposed to the air.

Miscibility with soil water and groundwater. Miscibility refers to how well a liquid pollutant can be mixed with and will remain mixed with water present under normal circumstances.

Solubility in soil water and groundwater. A pollutant's solubility is its ability to mix with water present under normal circumstances.

Density or specific gravity. Density is the ratio of the weight of a mass to the unit of volume. Specific gravity is the ratio of the weight of the volume of liquid or solid pollutant to the weight of an equal volume of water. Specific gravity of water is 1.0, so pollutants with a specific gravity below 1.0 will float on water and those with a specific gravity over 1.0 will sink in water.

Dynamic viscosity. This is the internal resistance of a liquid or gaseous pollutant while moving.

Reactivity. A pollutant's reactivity is the degree of its ability to undergo a chemical combination with another substance.

Susceptibility to biodegradation. This refers to the ease with which a pollutant breaks down into basic elements.

Along with the various pollutant properties that affect the fate of pollutants when they somehow enter soil, properties of the soil environment into which the pollutant is discharged are also important and need to be determined, especially when remediation activities are involved. These important properties, with respect to the fate of pollutants in soil, include the soil's:

- Infiltration capacity
- Natural organic content
- Saturated and unsaturated hydraulic conductivities
- Effective permeabilities for immiscible pollutants and soil water
- Mineralogy
- Oxygen content
- Bacterial community

Environmental practitioners whose job it is to protect soils or to remediate soil pollution must understand the several processes that work to control the rate and extent of migration of pollution in soils. These natural processes trap the pollutant, or delay the pollutant from spreading, and cause the pollutant to degrade or change in chemistry to a less hazardous state.

The processes that work to retard pollutant movement in soils include adsorption, ion (balance) exchange, chemical (pollutant) precipitation, and biodegradation.

Adsorption is the process by which one substance is attracted to and adheres to the surface of another substance without actually penetrating its internal structure. In soil, adsorption works to bond (hold) a pollutant to the surface of a soil particle or mineral in such a way that the substance is only available or disperses slowly. Clay and highly organic materials, for example, tend to adsorb pesticides rather than absorb them. Pollutant adsorption takes place on mineral surfaces where defects in their crystalline structure result in imbalances of electrical charges on the mineral surface. Dissolved pollutant molecules and ions with charge imbalances are attracted to mineral surfaces that have an opposite charge imbalance.

Ion balance is actually another step in the adsorption process whereby a dissolved pollutant substitutes itself for another chemical already adsorbed on the mineral surface.

Pollutant precipitation occurs when dissolved pollutants are removed (transported) from soil water and groundwater by various precipitation reactions. Chemical precipitation is also used as a pollution remediation technique whereby the pollutant is precipitated out of a contaminated stream.

Biodegradation is the breakdown of organic and pollutant compounds through microbial activity; it affects the distribution, movement (transport), and concentration of pollutants in soil.

The fate of certain pollutants applied to the soil surface or introduced into the soil profile depends on an extremely complex combination of interactive processes. The effectiveness of these interactive processes depends on several variable conditions—the nature of the pollutant, the manner of its application, the fundamental nature of the soil, and its transient state at the time and place of interest.

In addition to the various properties of pollutants and of the soil environment that affect pollutant migration, retardation, and transformation of mobile pollutants in soil, certain properties and/or conditions of the soil also affect the flow regime. These properties and conditions are the following:[8]

Soil texture. The migration of pollutants in coarse (sandy and gravelly) soil is generally faster than in fine (clay or silty) soils, which are more likely to retain pollutants and prevent their migration.

Layered soils (with vertical nonuniformity). These are more likely to retard migration in the soil profile than are uniform profiles.

Configuration of the soil layers. Whether soil layers are horizontal, slanted, or sloped has an effect on the pollutant flow regime. For example, if horizontal layers with concave troughs are present, pollutant flow is more likely to be retarded. On the other hand, in slanted or sloped layers, the pollutant's flow regime may not only be easier but may also be directed (for instance) toward a well or groundwater.

Depth of the water table. Obviously, if the depth to the water table is extreme, it is more difficult (in many cases, depending on the retardivity of the soil) for a pollutant to travel the entire distance through the soil to the water table.

Structure of the soil. Natural fissures, cracks, or channels in the soil are important considerations in identifying soil pathways, which enable rapid migration of pollutants in the soils.

Unstable flow. Pollutant flow not only migrates quickly through cracks and fissures but also has the tendency to concentrate in tongue-like streams (fingers), which generally begin at the transition from fine-textured to coarse-textured layers, bypassing or short-circuiting the greater volume of soil material and allowing direct transmission of pollutants to the water table.

Soil moisture. The soil's moisture content has an obvious effect on the pattern and migration of organic pollutants in the soil.

Origins of Soil Pollutants

All human societies, to some extent, pollute both soil and groundwater. People have had to develop ways of recognizing surface-water contamination or face disease and death from contaminated water throughout human history. Pollutants were often readily apparent and caused immediate, significant problems. However, contamination of the soil and the underground environment has remained virtually unnoticed until the last few decades because of its unseen locations and seemingly minimal impact.

We quite simply did not comprehend how effectively the mechanisms that carry contaminants through the soil could work, or the damage pollutants could do to the soil medium and the groundwater under soil's "protective" surface.

The number of human activities that cause underground contamination is much larger than most environmental scientists would have guessed even a few years ago. Soil quality problems originating on the surface include natural atmospheric deposition of gaseous and airborne particulate pollutants; infiltration of contaminated surface water; land disposal of solid and liquid waste materials; stockpiles, tailings, and spoil; dumps, salt spreading on roads; animal feedlots; fertilizers and pesticides; accidental spills; and composting of leaves and other yard wastes.

Other sources of soil contamination are related to petroleum products. These other sources include direct disposal of used oils on the ground by individuals or industries; seepage from landfills, illegal dumps, unlined pits, ponds, and lagoons; and spills from transport accidents or even auto accidents.[9]

GASEOUS AND AIRBORNE PARTICULATE POLLUTANTS

The following discussion focuses on contamination originating on the land surface. However, soil and subsurface contamination may also originate below ground (but above the water table) from septic tanks, landfills, sumps and dry wells, graveyards, USTs, underground pipeline leakages, and other sources. In addition, the soil, subsurface, and groundwater contamination may also originate below the water table from mines, test holes, agricultural drainage wells and canals, and others.

Soil figures prominently in the function of the carbon, nitrogen, and sulfur cycles—the biogeochemical cycles. While prominent in the normal operation of these cycles, soil also interfaces in powerful and essential ways with the atmosphere. Consider the nitrogen cycle, where nitrates and ammonium ions in rain water are absorbed by plant roots and soil microorganisms and converted to amino acids or to gaseous N_2 and N_2O, which diffuse back to the atmosphere. The N_2 uptake and conversion to amino acids (nitrogen fixation) by symbiotic and free-living soil microorganisms balances this loss of gaseous nitrogen. NO, NO_2, and NH_3 (other nitrogen gases) are also emitted and absorbed by soils. Soil reactions are major determinants of trace gas concentrations in the atmosphere.

Air pollutants—sulfur dioxide, hydrogen sulfide, hydrocarbons, carbon monoxide, ozone, and atmospheric nitrogen gases—are absorbed by soil. Because soil reactions are subtle, they are often ignored in tracing the effects of air pollution. However, two classic examples of airborne particulate soil contaminants can be seen in the accumulation of heavy metals around smelters and in soils in urban areas contaminated by auto exhaust emissions. While these two soil polluters can be serious in localized areas, long-range effects of such contamination are considered minor.

INFILTRATION OF CONTAMINATED SURFACE WATER

When a well is installed near a stream or river, the well's presence induces recharge from the water body, providing high yield with low drawdown. However, if the water body the well draws from is polluted, the soil water in the well field can become contaminated. This most commonly occurs when a shallow well draws water from the alluvial aquifer adjacent to the stream. The cone of depression imposed by pumping the well or well field creates a gradient on the water table directed toward the well. This pulls the polluted water through the soil and contaminates both the well field and the well.

LAND DISPOSAL OF SOLID AND LIQUID WASTE MATERIALS

Common practices for dealing with certain recyclable wastes (liquid and sludge [biosolids] wastes from sewage treatment plants, food processing companies, and other sources) include land disposal, stockpiling, or land-applying wastes or materials. These practices serve as both a means of disposal and a beneficial use/reuse of such materials. Such waste products often work successfully to fertilize agricultural lands, golf courses, city parks, and other areas, but the land selected for use must be carefully chosen and tested. Contamination problems arise if water-soluble, mobile wastes are carried deep into the subsurface. If water tables are near the surface, groundwater contamination problems can occur.

STOCKPILES, TAILINGS, AND SPOILS

The practice of stockpiling chemical products (if not properly managed and contained) contributes to soil and subsurface pollution. Road salt stockpiles can leach into the soil. Tailings produced in mining activities commonly contain materials (asbestos, arsenic, lead, and radioactive substances) that are a health threat to humans and other organisms. Tailing may also contain contaminants such as sulfide, which forms sulfuric acid upon contact with precipitation. The precipitation runs off or is leached from tailing piles, infiltrating the surface layer and contaminating soil. It may ultimately reach groundwater. Spoil (a common result of excavations where huge amounts of surface cover are removed, piled, and then moved somewhere else) causes problems similar to tailing problems; precipitation removes materials in solutions by percolating waters (leaching). Pollutants migrate from the spoil, finding their way into the soil and thus into shallow aquifers.

DUMPS

Illegal dumping is less common today than in the past, fortunately. In fact, uncontrolled dumping is now prohibited in most industrialized countries. The persistent remains of dumping are still with us, however. Dumping sites can contain just about anything. Such sites present localized threats of subsurface contamination.

SALT SPREADING ON ROADS

The practice of spreading deicing salts on highways is widespread, especially in urban areas in the North. Not only does this practice contribute to the deterioration of automobiles, bridges, and the roadway itself, it adversely affects plants growing alongside a treated highway or sidewalk. More seriously, salt contamination quickly leaches below the land surface. The productivity of the land decreases because most plants cannot grow in salty soils. Contamination of wells used for drinking water can occur in areas with long-term use.

ANIMAL FEEDLOTS

Some of the largest contributors to non-point-source water pollution are animal feed-lots. Animal waste in feedlots literally piles up. These stationary heaps, sometimes left for extended periods, create problems with runoff containing contaminants. These contaminants may not only enter the nearest surface water body but may contaminate soil by waste seepage.

FERTILIZERS AND PESTICIDES

The mainstays of high-yield agriculture are fertilizers and pesticides. The impact of these two practices on the environment has been significant, with each yielding different types of contaminants.

Most people think of the application of fertilizers and pesticides to soil as *treating* the soil. But is it treating it—or poisoning it? Environmental scientists are still trying to definitively answer this question. With fertilizer and pesticide application and its long-term effects, we do not know what we do not know—there is no proof of the long-term effects. The impact of using these chemicals is only now becoming clear.

ACCIDENTAL SPILLS

Disturbingly common, accidental spills of chemical products can be extremely damaging to any of the three environmental media—air, water, and soil. When chemical spills in the soil are not discovered right away, the contaminant may migrate through the soil (contaminating it) to the water table. As a general rule of thumb, we may assume that the impact of a chemical spill is directly related to the concentration at the point and time of release, the extent to which the concentration increases or decreases during exposure, and the time over which the exposure continues.

COMPOSTING OF LEAVES AND OTHER WASTES

Composting is a common practice for many homeowners who use the contained and controlled decay of yard and vegetable wastes as an environmentally friendly way to dispose of or beneficially reuse them. When the leaves, twigs, and other organic materials have been treated with chemical pesticides and some fertilizers, however, composting this may be harmful to the soil.

Soil Pollution Remediation

The rapid expansion and increasing sophistication of the chemical industries in the past century and particularly over the last thirty years has meant that there has been an increasing amount and complexity of toxic waste effluents. At the same time, fortunately, regulatory authorities have been paying more attention to problems of contamination of the environment.

The occurrence of major incidents (such as the *Exxon Valdez* oil spill, the Union-Carbide Bhopal disaster, large-scale contamination of the Rhine River, the progressive deterioration of the aquatic habitats and conifer forests in the northeastern United States, Canada, and parts of Europe, or the release of radioactive material in the Chernobyl accident, etc.) and the subsequent massive publicity due to the resulting environmental problems have highlighted the potential for imminent and long-term disasters in the public's conscience.

When chemical spills or accidents occur and soil is contaminated, "remediation" is the buzzword for mitigating the contamination problem. However, soil or subsurface remediation has its own problems; soil remediation technologies are not perfect resolutions, although they are often very effective. The first problem concerns the fact that soil pollution remediation practices are a still developing branch of environmental science and engineering—lessons are still being learned. Another more obvious problem has to do with timing and expediency. Remediation is the very last procedure in the sequence of combating soil pollution because it cannot begin until somebody realizes that the problem exists—after the spill or release of pollutants into the soil. Damage from water and air pollution can often be limited or prevented from occurring at all if the pollutant is collected at the source, before it enters the media. In soil pollution, often the damage is not only done, it's years old, and presents even more challenges to our ability to remediate the contamination.

The available technologies for remediating petroleum-product-contaminated soil and groundwater are divided into two categories: *in situ* treatment and non–*in situ* treatment—the treatment of soil in place and treatment of soil removed from the site, respectively. A compilation of various soil remediation technologies is presented in table 10.3.

CERCLA/RCRA

Because of the regulatory programs of CERCLA (Comprehensive Environmental Response, Compensation, and Liabilities Act of 1980—better known as *Superfund*) and RCRA (Resource Conservation and Recovery Act of 1976—better known as the *cradle to grave act*), common and widespread use of remediation is now included in the environmental vocabulary. How common is usage of the term *remediation*? To best answer this question, follow the response of venture capitalists in their attempts to gain a foothold in this new technological field. Regarding this issue, J. A. MacDonald points out: "In the early 1990s, venture capitalists began to flock to the market for

Table 10.3.   Soil remediation technologies.

Technology	Applicable Contaminants
In situ	
Natural attenuation	Petroleum hydrocarbons
	Chlorinated solvents
Passive	Petroleum hydrocarbons
	Chlorinated solvents
	Coal-tar residues
Leaching/chemical reaction	Petroleum hydrocarbons
	Chlorinated solvents
Isolation/containment	Petroleum hydrocarbons
	Chlorinated solvents
	Coal-tar residues
Stabilization	Petroleum hydrocarbons
	Chlorinated solvents
	Metals
	Coal-tar residues
Vitrification	Petroleum hydrocarbons
	Chlorinated solvents
	Coal-tar residues
Volatilization (vapor extraction)	Volatile organic compounds
Bioremediation[a]	Petroleum hydrocarbons[b]
Not in situ	
Land treatment	Petroleum hydrocarbons[c]
	Coal-tar residues
Thermal treatment	Petroleum hydrocarbons
	Chlorinated solvents
	Coal-tar residues
Solidification/stabilization	Petroleum hydrocarbons
	Chlorinated solvents
	Coal-tar residues
	Metals
Chemical extraction	Petroleum hydrocarbons
	Chlorinated solvents
	Coal-tar residues
	Metals
Excavation/disposal	Petroleum hydrocarbons
	Coal-tar residues
	Metals

Source: S. M. Testa, *The Reuse and Recycling of Contaminated Soil* (Boca Raton, Fla.: CRC/Lewis Publishers, 1997), 18.

Notes: a) Bioremediation refers to biotechnology whereby bacteria are altered to produce certain enzymes that metabolize industrial waste components that are toxic to other life, and also new pathways are designed for the biodegradation of various wastes; b) Excludes VOCs; c) Petroleum hydrocarbons include gasoline and fuel oils such as diesel and kerosene.

groundwater and soil cleanup technologies, seeing it as offering significant new profit potential. The market appeared large; not only was $9 billion per year being spent on contaminated site cleanup, but existing technologies were incapable of remediating many serious contamination problems."[10]

✔ **Key Point:** From MacDonald's comments, soil remediation technology appears a booming enterprise with unlimited potential, but inherent problems can limit this potential. Remediation technology is a double-edged sword.

Numerous remediation technologies (also commonly known as "innovative cleanup technologies") were developed and became commercially available after RCRA and CERCLA were put into place. We have briefly discussed many of these technologies.

Although the goals of regulatory agencies monitoring environmental cleanup efforts are generally going to include removing every trace of contamination and restoring the landscape to its natural condition, achieving that end result is highly unlikely.

Additional Reading

Bohn, H. L. L., et al. *Soil Chemistry*. New York: John Wiley & Sons, 2001.
Orlov, D. S. *Soil Chemistry*. Russia: Balkem, A.A., 1992.
Sparks, D. L., *Soil Physical Chemistry*. 2nd ed. Boca Raton, Fla.: CRC Press, 1998.

Summary

- Soil contamination is a major environmental concern not only throughout the United States but worldwide.
- Regardless of the origin, most soils consist of four basic components: mineral matter, water, air, and organic matter.
- Petroleum contamination is commonly associated with USTs.
- Besides the obvious problem of fouling the environment, many of these leaking USTs also pose serious fire and explosion hazards.
- Experience has shown that it is relatively easy to describe the fate of petroleum in soils in merely qualitative terms.
- Geothermal energy is used best in the generation of electricity.
- Manufactured-gas plants produce a variety of largely hazardous waste products.
- Mining operations often create land and water pollution problems.
- Cation-exchange capacity is the soil's capacity to hold cations.
- The pH of soil is one of the most important properties involved in plant growth.
- Saline soils are soils with lots of soluble salts.
- Soil color is dependent primarily on the quantity of humus and the chemical form of iron oxides present.
- Soil texture refers to the relative proportions of sand, silt, and clay in a soil.

- Soil tilth refers to the combination or arrangement of primary soil particles into secondary particles.
- About half of soil's volume is pore space, which is occupied by varying amounts of air and water, depending on how wet the soil is.
- Drainage refers to a soil's ability to get rid of excess water.
- Soil depth refers to how deep, top to bottom, the topsoil plus the subsoil is.
- The slope has a marked influence on the amount of water runoff and soil erosion caused by flowing water.
- It is important to understand soil's transport mechanisms affecting the flow regime of contaminants or pollutants.
- Pollutant retardants include adsorption, ion balance, pollutant precipitation, and biodegradation.
- All human societies, to some extent, pollute both soil and groundwater.
- When chemical spills or accidents occur and soil is contaminated, "remediation" is the buzzword for mitigating the contamination problem.
- Although the goals of regulatory agencies monitoring environmental cleanup efforts are generally going to include removing every trace of contamination and restoring the landscape to its natural condition, achieving that end result is highly unlikely.

New Word Review

Absorption—the process by which one substance is taken into and included with another substance.

Acid soil—soil with a pH value < 7.0.

Adsorption—the increased concentration of molecules or ions at a surface, including exchangeable cations and ions on soil particles.

Aggregate—a unit of soil structure, usually formed by natural processes in contrast with artificial processes and generally <10 mm in diameter.

Air porosity—the fraction of the bulk volume of soil that is filled with air at any given time or under a given condition, such as a specified soil-water content or soil-water matrix potential.

Available water—the portion of water in a soil that can be readily absorbed by plant roots. Considered by most workers to be that water held in the soil against a pressure up to approximately 15 bars.

Cation-exchange Capacity (CEC)—the sum total of exchangeable cations that a soil can absorb.

Comprehensive Environmental Response, Compensation, and Liability Act (CERCLA)—the statute, also called the Superfund, establishes federal authority for emergency response and cleanup of hazardous substances that have been spilled, improperly disposed of, or released into the environment. It was enacted in 1980 and significantly amended in 1984 and in 1986 through the Superfund Amendments and Reauthorization Act (SARA). The primary responsibility for cleanup rests with the generator or disposer of the hazardous substance with the back-up federal response using a trust fund.

Humus—total of the organic compounds in soil exclusive of undecayed plant and animal tissues, their "partial decomposition" products, and the soil biomass.

Leaching—the removal of materials in solution from the soil.

Parent material—the unconsolidated and more or less chemically weathered mineral or organic matter form which the solum of soil is developed by pedogenic processes.

Resource Conservation and Recovery Act (RCRA)—the federal statute of 1976, amended in 1980, and 1984, which itself was an amendment of the Solid Waste Disposal Act of 1965. It provides for the management of solid (nonhazardous) and hazardous wastes. EPA has set minimum standards for all waste disposal facilities. For hazardous wastes, it has set standards for generation, transportation, storage, and treatment of wastes. The underground storage program and medical waste tracking have been added.

Salinization—the process of accumulation of salts in soil.

Soil—the unconsolidated mineral or material on the immediate surface of the earth that serves as a natural medium for the growth of land plants.

Tilth—the physical condition of soil as related to its ease of tillage, fitness as a seedbed, and is impedance to seedling emergence and root penetration.

Chapter Review Questions

10.1. _____ affects the largest number of sites and the largest total volume of impacted material.

10.2. _____ are the bridge between mineral matter and life.

10.3. One of the biggest problems in using geothermal energy is _____.

10.4. Manufactured gas plants produce _____ waste.

10.5. _____ releases waste materials that allow environmental poisons entry into the soil.

10.6. As organic matter decomposes in a soil, _____ are produced.

10.7. Soil pH can be adjusted by _____.

10.8. The _____ of a soil can give clues to its health, origin, and long-term changes.

10.9. _____ refers to a soil's ability to get rid of excess water.

10.10. _____ is the property that determines how easily a pollutant evaporates.

10.11. The breakdown of organic and pollutant compounds through microbial activity is called _____.

10.12. Some of the largest contributors to non-point-source water pollution are _____.

Notes

1. H. C. Brookfield, "Sensitivity to Global Change: A New Task for Old/New Geographers." Norma Wilkinson Memorial Lecture, University of Reading. UK, 1989.

2. D. L. Winegardner, *An Introduction to Soils for Environmental Professionals* (Boca Raton, Fla.: CRC/Lewis Publishers, 1996).

3. S. M. Testa, *The Reuse and Recycling of Contaminated Soil* (Boca Raton, Fla.: CRC/Lewis Publishers, 1997).

4. Testa, *Reuse and Recycling*, 6.

5. L. Eastcott, W. Y. Shir, and D. Mackay, "Modeling Petroleum Products in Soil, in *Petroleum Contaminated Soils*, vol. 1, ed. Paul T. Kostecki and Edward J. Calabrese (Chelsea, Mich.: Lewis Publishers, 1989).

6. M. M. Reed, "Environmental Compatibility of Geothermal Energy." U.S. Department of Energy Technical Site, http://geotherm.inel.gov/geothermal/artciles:reed/index.html (accessed December 25, 2002).

7. "What is Geothermal Energy, Geothermal Technologies," http://www.eren.doe.goc/geothermal/about.html.

8. P. T. Kostecki and E. J. Calabrese, *Petroleum Contaminated Soils*, vol. 1, *Remediation Techniques, Environmental Fate, and Risk Assessment* (Chelsea, Mich.: Lewis Publishers, 1989).

9. R. K. Tucker, "Problems Dealing with Petroleum Contaminated Soils: A New Jersey Perspective," in Kostecki and Calabrese, eds., *Petroleum Contaminated Soils,* vol. 1.

10. J. A. MacDonald, "Hard Times for Innovation Cleanup Technology," *Environmental Science and Technology* 31, no. 12 (December 1997): 560–63.

CHAPTER 11

Organic Chemistry and Pesticides

Black and portentous must this humour prove,
Unless good counsel may the cause remove.

—Shakespeare, *Romeo and Juliet*, I, i

Topics in This Chapter

- What Is a Pesticide?
- Environmental Practitioners and Pesticide Control
- Nitrogen Chemistry
- Phosphorus Chemistry
- Sulfur Chemistry
- Pesticide Plan

When pesticide is applied, what happens to it? Once they reach the soil, organic chemicals such as pesticides "move in one or more of seven directions: (1) They may vaporize into the atmosphere without chemical change; (2) they may be absorbed by soils; (3) they may move downward through the soil in liquid or solution form and be lost from the soil by leaching; (4) they may undergo chemical reactions within or on the surface of the soil; (5) they may be broken down by soil microorganisms; (6) they may wash into streams and rivers in surface runoff; and (7) they may be taken up by plants or soil animals and move up the food chain."[1]

Pesticide losses depend primarily on the intensity and duration of rainfall, the time between herbicide application and the next runoff event, temperature, sunlight intensity, wind speed and duration, concentration of the pesticide in the soil, chemical properties of the pesticide, tillage practice, soil type, pH, organic matter, and slope.

✔ **Key Point:** Pesticides are released deliberately into the air, water, and soil, not incidentally, but they have many incidental effects. If they are aimed at pests, they hit the environment generally. In the study of pesticides, it is important for the reader to ask a broad question—that of the gap between narrowly focused intentions and widely dispersed effects: "How could intelligent beings seek to control a few unwanted species by a method that contaminated the entire environment?"[2]

Information in this section is from USEPA's *What Is a Pesticide?* http://www.epa.gov/opp00001/whatis.htm (accessed December 26, 2002).

225

What Is a Pesticide?

A *pesticide* is any substance or mixture of substances intended for preventing, destroying, repelling, or mitigating any pest. Pests can be insects, mice and other animals, unwanted plants (weeds), fungi, or microorganisms like bacteria and viruses. Though often misunderstood to refer only to *insecticides,* the term also applies to herbicides, fungicides, and various other substances used to control pests. Under United States law, a pesticide is also any substance or mixture of substances intended for use as plant regulator, defoliant, or desiccant.

Many household products are pesticides. Did you know that all of the following common products are considered pesticides?

- Cockroach sprays and baits
- Insect repellents for personal use
- Rat and other rodent poisons
- Flea and tick sprays, powders, and pet collars
- Kitchen, laundry, and bath disinfectants and sanitizers
- Products that kill mold and mildew
- Some lawn and garden products, such as weed killers
- Some swimming pool chemicals

By their very nature, most pesticides create some risk of harm to humans, animals, or the environment because they are designed to kill or otherwise adversely affect living organisms. At the same time, pesticides are useful to society because of their ability to kill potential disease-causing organisms and control insects, weeds, and other pests.

✔ **Key Point:** In the United States, the Office of Pesticide Programs of the USEPA is chiefly responsible for regulating pesticides. Biologically based pesticides, such as pheromones and microbial pesticides, are becoming increasingly popular and often are safer than traditional chemical pesticides.

Some common kinds of pesticides and their function are presented in table 11.1.

✔ **Key Point:** The term pesticide also includes these substances: Defoliants, desiccants, insect growth regulators, and plant growth regulators.

Because the U.S. definition of pesticides is quite broad, USEPA also defines what is *not* a pesticide.

- Drugs used to control diseases of humans or animals (such as livestock and pets) are not considered pesticides; such drugs are regulated by the Food and Drug Administration (FDA).
- Fertilizers, nutrients, and other substances used to promote plant survival and health are not considered plant growth regulators and thus are not pesticides.

Table 11.1. Pesticides and their functions.

Pesticides	Function
Algicides	Control algae in lakes, canals, swimming pools, water tanks, and other sites
Antifouling agents	Kill or repel organisms that attach to underwater surfaces, such as boat bottoms
Antimicrobials	Kill microorganisms (such as bacteria and viruses)
Attractants	Attract pests (lure an insect or rodent to a trap)
Biocides	Kill microorganisms
Disinfectants and sanitizers	Kill or inactivate disease-producing microorganisms on inanimate objects
Fumigants	Produce gas or vapor intended to destroy pests in buildings or soil
Fungicides	Kill fungi (including blights, mildews, molds, and rusts)
Herbicides	Kill weeds and other plants that grow where they are not wanted
Insecticides	Kill insects and other arthropods
Microbial pesticides	Kill, inhibit, or out compete pests, including insects or other microorganisms
Miticides	Kill mites that feed on plants and animals
Molluscicides	Kill snails and slugs
Nematicides	Kill nematodes (microscopic, worm-like organisms that feed on plant roots)
Ovicides	Kill eggs of insects and mites
Pheromones	Biochemicals used to disrupt the mating behavior of insects
Repellents	Repel pests, including insects (such as mosquitoes) and birds
Rodenticides	Control mice and other rodents

- EPA exempts biological control agents, except for certain microorganisms, from regulation.
- Products that contain certain low-risk ingredients, such as garlic and mint oil, have been exempted from federal registration requirement, although state regulatory requirements may still apply.

✔ **Important Point:** For a list of ingredients that may be exempt, and a discussion of allowable label claims for such products, see USEPA's Pesticide Registration Notice 1000-6, "Minimum Risk Pesticides Exempted under FIFRA Section 25(b)."

Environmental Practitioners and Pesticide Control

One of the major goals of the environmental practitioner is to promote safe work practices in an effort to minimize the incidence of illness and injury experienced by employees. Though stated as one major goal, it should be obvious that actual accomplishment requires a wide range of interrelated factors. The environmental professional does not necessarily have to be an "expert" in any one of the numerous factors involved but instead must be a generalist with knowledge in all pertinent areas.

Effective pesticide control in the workplace requires not only knowledge of pesticides, but also employee training. In addition, to be effective, an organization's control of pesticides in the workplace must be guided by a formal *pesticide plan*. In many workplaces, the responsible person in charge of the pesticide plan is the environmental professional.

Simply, the environmental professional must not only be well grounded in the chemistry of pesticides, and in pesticides themselves, but also must act as the administrator of the organization's pesticide control plan.

🖊 **Key Point:** Experience has shown that an environmental professional's assignment as "administrator" of a safety and health program does not mean that he or she is the "enforcer" of such programs. A better choice of words and actions would be an "advocate" for such compliance.

In the sections that follow, we concentrate on two main areas related to pesticides: The chemistry of nitrogen, phosphorus, and sulfur (especially as they occur in pesticides, herbicides, and insecticides, medicines, and key industrial compounds) and a sample pesticide plan.

Nitrogen Chemistry

Nitrogen (N) makes up about 78% of Earth's atmosphere by volume (the atmosphere of Mars, in contrast, contains less than 3% nitrogen). However, its compounds are vital components of foods, fertilizers, and explosives. Nitrogen gas is colorless, odorless, and generally inert. As a liquid it is also colorless and odorless. The element seemed so inert that Lavoisier named it *azote*, meaning "without life."

When nitrogen is heated, it combines directly with magnesium, lithium, or calcium. When mixed with oxygen and subjected to electric sparks, it forms nitric oxide (NO) and then the dioxide (NO_2). When heated under pressure with hydrogen in the presence of a suitable catalyst, it forms ammonia. Nitrogen is "fixed" from the atmosphere by bacteria in the roots of certain plants such as clover—hence the usefulness of clover in crop rotation.

Apart from the atmosphere, nitrogen naturally occurs in soils via the decomposition of animal matter by bacteria, manures (hence their value as fertilizers), and plant and animal proteins.

Despite its inability to support life, nitrogen, as mentioned, is essential to many living organisms. For example, nitrogen makes up about 3% of our body weight, which we obtain from plant and animal *proteins*. Proteins are made up of *amino acids* (see figure 11.1).

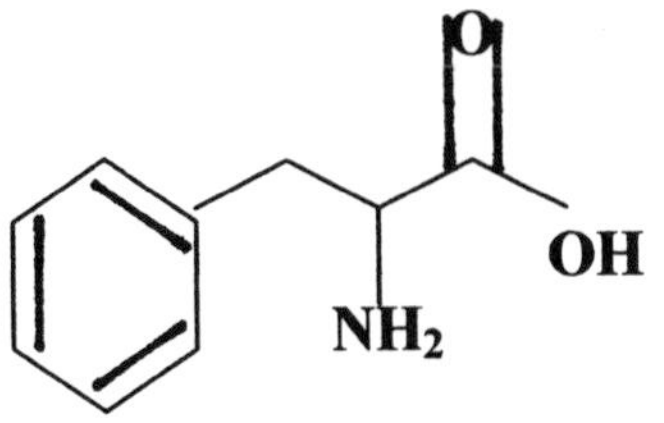

Phenylalanine

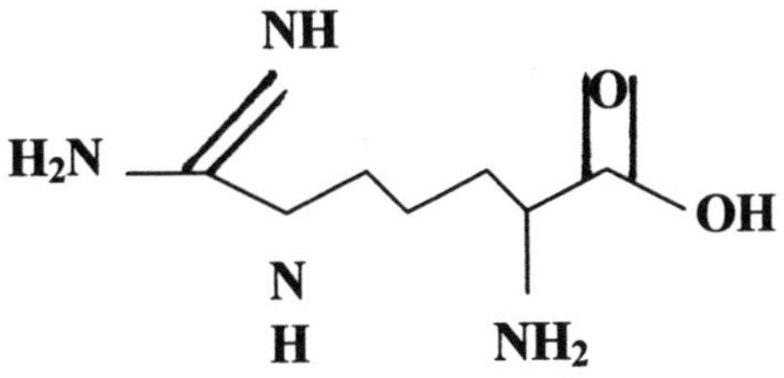

Arginine

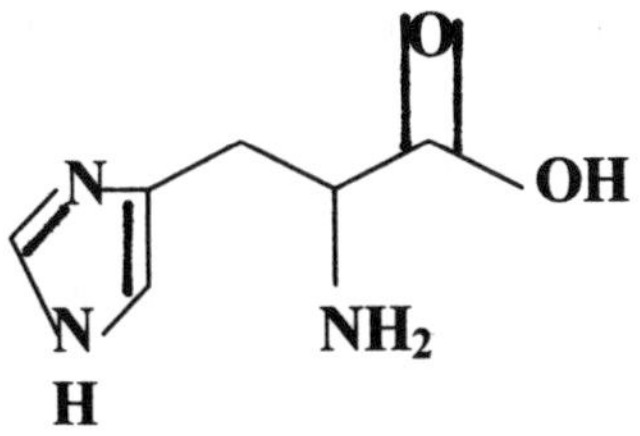

Histidine

Figure 11.1.   Amino acids.

Atmospheric nitrogen (N_2) is remarkably inert. The triple bond between the two nitrogens must be broken before nitrogen can react (see figure 11.2).

Because diatomic nitrogen is relatively unreactive, very little nitrogen is available in a form that can be used by living organisms. Therefore we rely on microorganisms to *fix* nitrogen into nitrate, which can be used by both plants and animals to make proteins.

Nitrogen fixation can result in a wide variety of compounds because nitrogen compounds have a large range of *oxidation states*.

$$N^{-3} \text{ (gains 3 electrons)} \leftrightarrow N^{+5} \text{ (loses 5 electrons)}$$

N≡N

1225 kcal/mol

(24% stronger)

C≡C

196 kcal/mol

Figure 11.2.   Nitrogen bond.

Important nitrogen compounds include:

- Amines
- Nitrogen oxides (NO$_x$)
- Explosives
- Carbamates and isocyanates
- Nucleic acids
- Urea and guanidine
- Optical bleaches

AMINES: AMMONIA (NH$_3$)

Ammonia (NH$_3$) is produced in nature when any nitrogen-containing organic material decomposes in the absence of air. It has a pungent odor detectable at approximately 50 to 60 ppm. It irritates the eyes and respiratory tract at concentrations as low as 100 to 200 ppm and is rapidly fatal at significantly higher concentrations by causing severe irritation to the lungs. Ammonia is produced on an industrial scale by reacting hydrogen gas and nitrogen gas.

$$N_2(g) + 3H_2(g) \rightarrow 2NH_3(g)$$

NH$_3$ is used in the production of fertilizers, nitric acid, explosives, drugs, and dyes. Liquid ammonia is also an important commercial refrigerant. It boils at −33.4°C and absorbs 322 calories/gram during evaporation. Only water has a higher heat of vaporization.

Ammonia is readily soluble in water, in part because of its ability to hydrogen bond (see figure 11.3).

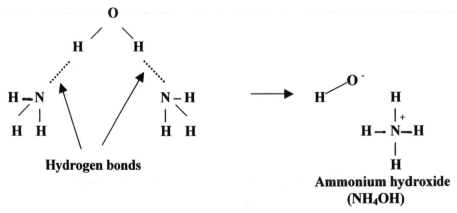

Figure 11.3. Ammonia solubility in water.

Alkaloids are naturally occurring amines, which have marked pharmacological properties at low doses (see figure 11.4).

NITROGEN OXIDES: NITRIC ACID (HNO₃)

Nitric acid was known to the alchemists of the eighth century as *aqua fortis* ("strong water"). It was prepared from KNO_3 and was used in the separation of gold from silver; it dissolves silver but leaves gold. Traces of nitric acid occur in the atmosphere after thunderstorms, and its salts are widely used in nature.

On the industrial scale, nitric acid is typically produced from ammonia in a three-step process.

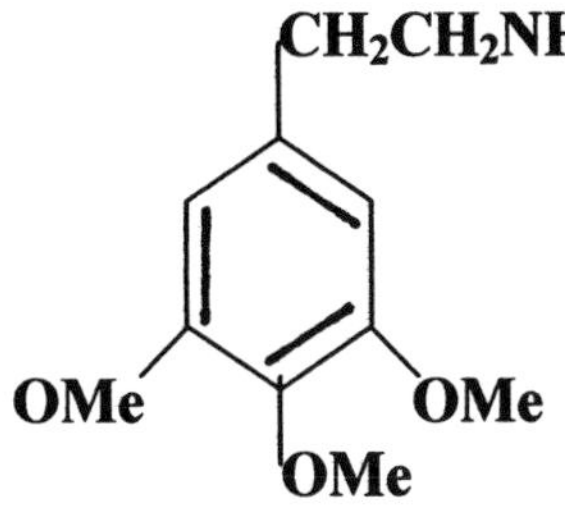

Mescalline
from mescal

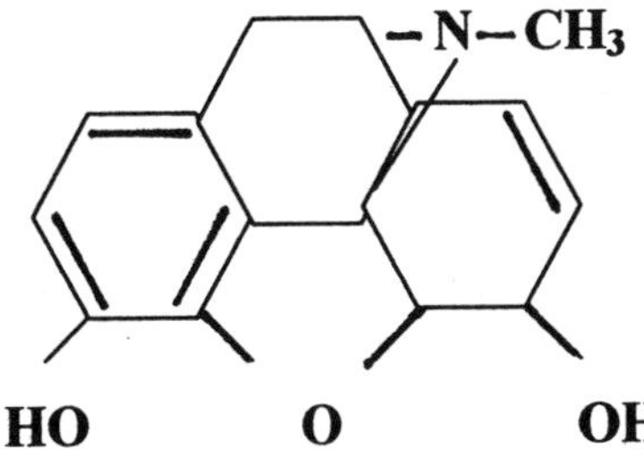

Morphine
from opium

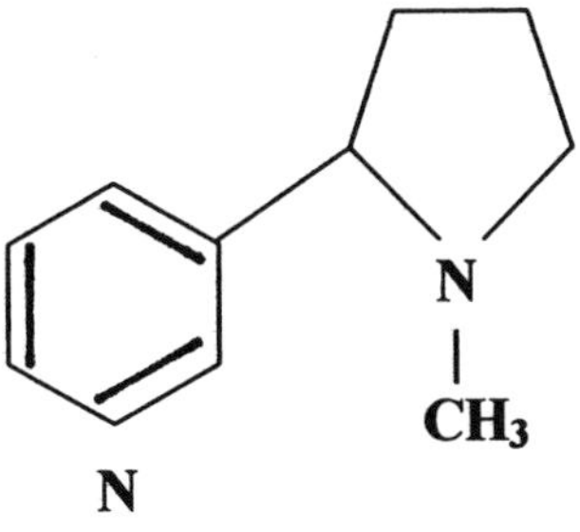

Nicotine
from tobacco

Figure 11.4. Amines with pharmacological properties.

Table 11.2. Other nitrogen oxides.

Oxide	Name(s)	Ox State	Color
N_2O	Nitrogen (I) oxide	+1	Colorless*
NO	Nitrogen (II) oxide	+2	Colorless
N_2O_3	Nitrogen (III) oxide	+3	Blue
NO_2	Nitrogen (IV) oxide	+4	Red-brown**
N_2O_4	Dinitrogen tetroxide	+4	Colorless
N_2O_5	Nitrogen (V) oxide	+5	White

* Laughing gas
** Smog

$$4NH_3(g) + 5O_2(g) \rightarrow 4NO(g) + 6H_2O(g)$$
$$2NO(g) + O_2(g) \rightarrow 2NO_2(g)$$
$$3NO_2(g) + H_2O(l) \rightarrow 2HNO_3(aq) + NO(g)$$

Approximately 75% of nitric acid is synthesized for the production of fertilizers (primarily ammonium nitrate, NH_4NO_3). An additional 15% of manufactured nitric acid is used in the production of explosives.

Nitric acid and nitrous acid (HNO_2) are also formed in the atmosphere in a complex series of reactions from nitrogen-based pollutants:

$$2NO(g) + H_2O(l) \rightarrow HNO_3(aq) + HNO_2(aq)$$
$$3HNO_2(aq) \rightarrow HNO_3(aq) + 2NO(g) + H_2O(l)$$
acid rain

Other nitrogen oxides are presented in table 11.2.

Ammonium nitrate is a well-known fertilizer that also has caused significant damage to lakes and streams due to *eutrophication*, leading to prolific growth of aquatic plants and shortening the life of the water body (see figure 11.5). Eutrophication is the process that results from accumulation of nutrients in lakes or other water bodies. It is a natural process, but can be greatly accelerated by human activities. Overgrowth of aquatic vegetation, in turn, affects dissolved oxygen, temperature, and other indicators.

✔ **Caution:** Overheated ammonium nitrate will explode.

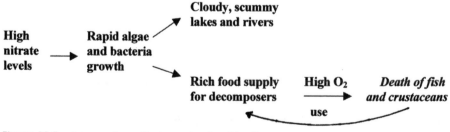

Figure 11.5. Ammonium nitrate and eutrophication.

Nitrates also contaminate drinking water supplies. Excess nitrates can cause hypoxia (low levels of dissolved oxygen—blue baby syndrome) and can become toxic to warm-blooded animals at higher concentrations (10 mg/L or higher) under certain conditions. The natural level of ammonia or nitrate in surface water is typically low (less than 1 mg/L); in the effluent of wastewater treatment plants, it can range up to 30 mg/L.

NO_3^- (natural bacteria in the gut) + NO_2^- (converts hemoglobin into forms that cannot carry O_2) = blue baby syndrome

Ammonium nitrate and other nitrogen compounds are also *explosive* (see figure 11.6).

CARBAMATES AND ISOCYANATES

Carbamate pesticides are commonly used in agriculture, as insecticides, fungicides, herbicides, nematocides, or sprout inhibitors. In addition, they are used as biocides for industrial or other applications and in household products. They are derived from a carbamic acid skeleton (see figure 11.7). More than fifty carbamates are known. Common carbamate pesticides include:

• *Carbaryl*, a lawn and garden insecticide, that has low toxicity in mammals
• *Carbofuran*, which kills insects that eat plants by disrupting acetylcholinesterase activity
• *Paraquat*, a highly toxic herbicide that rapidly disrupts enzyme and organ function, including the liver, heart, and kidneys

✔ **Key Point:** Carbonyl is a carbamate pesticide sold under the trade name Sevin.

Isocyanates ($R - N = C = O$) include *methyl isocyanate* (MIC), which is best known as the chemical agent responsible for the infamous Bhopal, India, incident in 1984—the

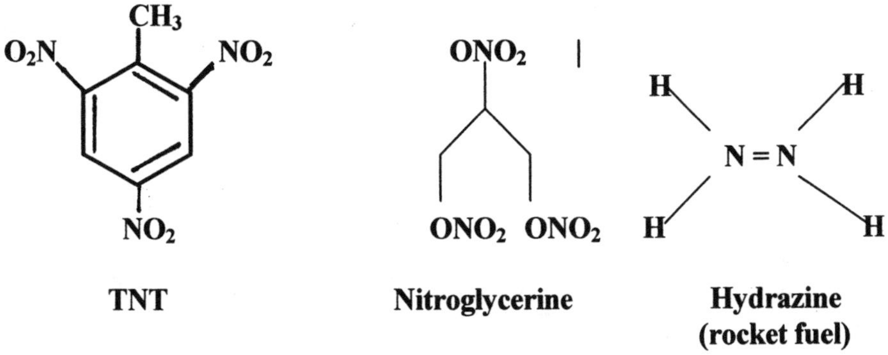

Figure 11.6. Nitrogen compound explosives.

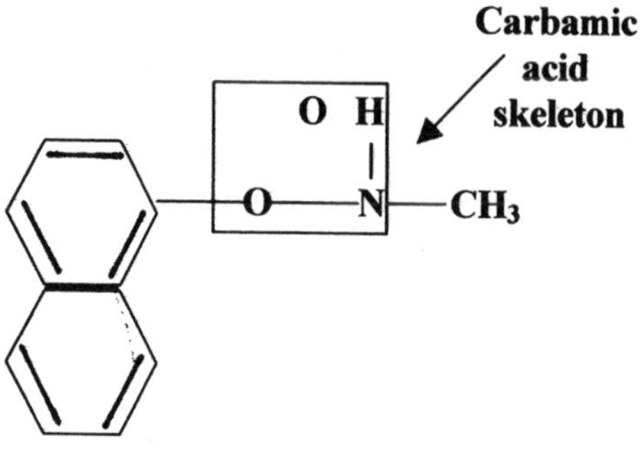

Carbaryl
(Sevin)

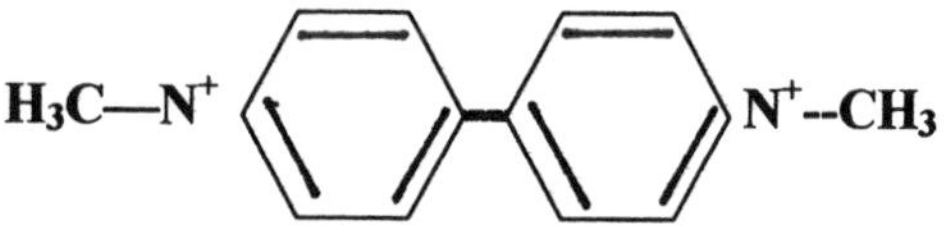

Paraquat

Figure 11.7. Two carbamate pesticides.

worst industrial accident in history. Methyl isocyanate is a volatile, colorless liquid with a strong, sharp odor that causes tears. Symptoms of poisoning include eye/nose/throat irritation, coughing, secretions, chest pain, dyspnea, asthma, and eye and skin injury. It is used in making pesticides, polyurethane foams, and plastics.

$$CH_3 - N = C = O$$

NUCLEIC ACIDS

The basis for understanding biochemical genetics requires understanding of the chemical nature of nucleic acids. The most important nucleic acids are macromolecules of very high molecular weights whose sole function is the storage and transfer of generic information. There are two types of such nucleic acids: *ribonucleic acids* (RNA), located primarily in the cytoplasm of the cells, and *deoxyribonucleic acids* (DNA), concentrated in the nucleus of the cell.

The primary structure of both DNA and RNA consists of an alternating sugar phosphate backbone linked to nitrogen heterocyclic compounds called *bases* (see figure 11.8).

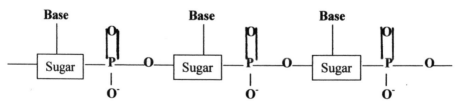

Figure 11.8. DNA and RNA bases.

The four DNA bases are shown in figure 11.9.

The four RNA bases include the three DNA bases, adenine, guanine, and cytosine (shown in figure 11.9). The fourth base, unique to RNA, is uracil (shown in figure 11.10).

UREA AND GUANIDINE

Lower animals excrete excess nitrogen from protein metabolism as ammonia, by higher animals as urea, and by birds and reptiles as *guanidine* (see figure 11.11).

OPTICAL BLEACHES

The so-called optical bleaches are not really bleaches but are fluorescent dyes that enhance the reflection of light. They absorb ultraviolet light and fluoresce blue light, which masks the natural yellowing of white fabrics.

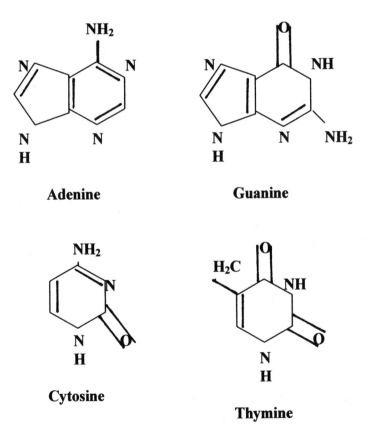

Adenine

Guanine

Cytosine

Thymine

Figure 11.9. The four DNA bases.

Adenine—DNA/RNA

Guanine—DNA/RNA

Cytosine—DNA/RNA

Uracil—RNA (only)

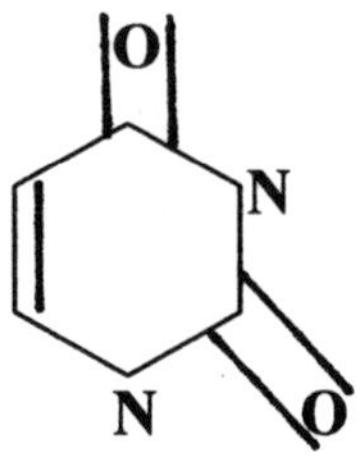

Uracil

Figure 11.10.   The four main RNA bases.

Urea   **Guanidine**

Figure 11.11.   Urea and guanidine.

Phosphorus Chemistry

Phosphorus, first discovered in 1669 by a chemist named Brand, comes from many sources, both natural and human. These include soil and rocks, wastewater treatment plants, runoff from fertilized lawns and cropland, failing septic systems, runoff from animal manure storage areas, disturbed land areas, drained wetlands, water treatment, and commercial cleaning preparations.

Phosphorus has a complicated story. Pure, "elemental" phosphorus (P) is rare; it is very reactive. In nature, phosphorus usually exists as part of a phosphate molecule

(PO$_4$). Phosphorus in aquatic systems occurs as organic phosphate and inorganic phosphate. Organic phosphate consists of a phosphate molecule associated with a carbon-based molecule, as in plant or animal tissue. Phosphate that is not associated with organic material is inorganic, the form required by plants. Animals can use either organic or inorganic phosphate.

✔ **Key Point:** Because phosphorus has 5 outer shell electrons, its most common oxidation states are -3 and $+5$.

There are several forms of phosphorus; two of the most common are referred to as *white phosphorus* and *red phosphorus*.

$$P(white) \rightarrow 4200 \text{ cal.} \rightarrow P(red)$$

A comparison of red and white phosphorus is presented in table 11.3.

Phosphorus was first used in matches in 1827. The original "strike anywhere" matches typically contained paraffin, 4%–7% white phosphorus (now P$_4$S$_3$), lead dioxide, and powdered glass.

✔ **Key Point:** Friction exploded the phosphorus/lead dioxide mixture, which in turn ignited the paraffin, which in turn ignited the wood. Modern safety matches contain no phosphorus, but only materials that will ignite when struck against the side of the box, impregnated with red phosphorus.

Phosphorus is also used in the manufacture of pyrotechnics, incendiary shells, smoke bombs, tracer bullets, fertilizers, special glasses, fine chinaware, cleaning agents, and pesticides. The primary industrial uses of phosphorus include fertilizers; animal feed supplements, detergents, pesticides, medicines, military poisons, and incendiary devices.

Complex phosphorus-containing compounds are also essential biological constituents. On average, a man's body contains significant amounts of phosphorus (skeleton, 1400g; muscles, 130g; and nerves/brain, 12g). Egg yolks, beans, nuts, peas, and wheat also contain phosphorus.

Because P is in the same group on the periodic table as N, there is a similarity between the formulas of N and P compounds. However, the *properties* of those com-

Table 11.3. Comparison of red and white phosphorus.

Properties	White	Red
H$_2$O Solubility	Nil	Nil
CS$_2$ Solubility	70g/100g (10°C)	Nil
MP	44.1°C	590°C
BP	280°C	Sublimes
Density g/cc	1.83	2.34
Formula	P$_4$	P$_4$

pounds are often considerably different. Table 11.4 presents a comparison of NH_3 and PH_3. Several important phosphorus compounds are:

- Phosphates
- Acids
- Halides
- Oxides
- Pesticides and other poisons
- Phospholipids and nucleic acids

PHOSPHATES

Phosphorus is commonly found in nature in the form of phosphates $(-PO_4^{-2})$. One of the primary natural sources of phosphates is from the *apatites,* a common mineral with the general formula $Ca_5(PO_4)_3(F, OH, Cl)$ that is found in the southern and western states (Florida, Tennessee, and Montana). Phosphates are also mined from marine sediments and guano.

▶ **Key Point:** Calcium phosphate is found in all fertile soils and constitutes approximately 60% of the bones and teeth of animals.

The United States produces about 40 million metric tons of phosphate each year and holds estimated reserves of 105 billion tons of phosphate rock (approximately 1.4 million tons of phosphate).

Commonly occurring *tricalcium phosphate* ("bone ash") is an amorphous, odorless, tasteless powder. It is used in the manufacture of mill-glass, polishing and dental powders, porcelains, pottery; enameling, and clarifying sugar syrups; in animal feeds; as noncaking agent; and in the textile industry. For medical use it is a calcium replacement and antacid.

Tricalcium phosphate is very insoluble in water and is converted to the more soluble *monocalcium phosphate* (calcium superphosphate) for fertilizer.

$$Ca_3(PO_4)_2 + 2H_2SO_4 \rightarrow Ca(H_2PO_4)_2 + 2CaSO_4 \text{ "superphosphate"}$$

Table 11.4. Comparison of NH_3 and PH_3.

Properties	NH_3	PH_3
BP	$-33.4°C$	$-87.7°C$
MP	$-77.7°C$	$-133°C$
Autoignition Temperature	nonflammable	$100°C$
Toxicity	eye/nose/throat irritation $LC_{50} = \sim8$ ppm	poisonous!
Other	——	violent reaction with O_2, halides

Condensed *polyphosphates* (condensed phosphates) are strong complexing agents for some metal ions. Polyphosphates are used for treating boiler waters and in detergents. In water, polyphosphates are unstable and will eventually convert to orthophosphate. Condensed polyphosphates such as sodium triphosphate are of great importance in the manufacture of detergents. Most laundry detergents contain approximately 35% to 75% triphosphate, which serves two purposes: it provides an alkaline solution (pH 9.0 to 10.5) necessary for effective cleaning and prevents interferance in the cleansing role of the detergent.

These condensed polyphosphates serve to adjust pH and complex water-hardening agents such as Ca^{+2} and Mg^{+2}. For example, Calgon is a high-molecular-weight phosphate (Graham's salt).

Phosphate detergents have also been implicated in eutrophication. Alga growth is limited by the available supply of phosphorus or nitrogen, so if excessive amounts of these nutrients are added to the water, algae and aquatic plants can grow in large quantities. When these algae die, they are decomposed by bacteria, which use dissolved oxygen. (the eutrophication process). Dissolved oxygen concentrations can drop too low for fish to breathe, leading to fish kills. Excessive amounts of algae grow into scum on the water surface, decreasing recreational value and clogging water-intake pipes. Rapid decomposition of dense alga scums with associated organisms can give rise to foul odors.

In freshwater lakes and rivers, phosphorus is often the growth-limiting nutrient, because it occurs in the least amount relative to the needs of plants. In estuaries and coastal waters, nitrogen is generally the growth-limiting nutrient.

🖎 **Key Point:** Because phosphate detergents have been implicated in eutrophication, they have consequently been replaced by carbonate and silicate salts.

Phosphates have important functions in households and in the economy:

- *Monocalcium phosphate* [$Ca(H_2PO_4)$] is an important water-soluble ingredient of fertilizer.
- *Monosodium phosphate* [NaH_2PO_4] is added to baking powder because of its acidic properties.
- *Disodium phosphate* [Na_2HPO_4] is a water softener which precipitates calcium.
- *Trisodium phosphate* [Na_3PO_4] is both a degreaser and a water-softening agent.

PHOSPHORUS ACIDS

There are many acids of phosphorus, but two of the most common are *phosphoric acid* (H_3PO_4) and *phosphorous acid* (H_3PO_3). The primary uses for these compounds are the manufacture of phosphates for fertilizers, trisodium phosphate, baking powder, and carbonated drinks and syrups.

PHOSPHORUS HALIDES

The two most important phosphorus halides are *phosphorus trichloride* (PCl_3) and *pentachloride* (PCl_5). These compounds are used as catalysts and chlorinating agents, as well as precursors to various oxyhalides, such as *phosphorus oxyhalide* ($POCl_3$).

OXIDES OF PHOSPHORUS

The two oxides of phosphorus (P_4O_6 and P_4O_{10}) are typically referred to by their *empirical formulas* as *tri*oxide and *pent*oxide, respectively. Both react readily with water to form the corresponding acids.

$$P_4O_6 \rightarrow (\text{cold water + agitation}) \rightarrow 4H_3PO_3$$
$$P_4O_{10} + 6H_2O \rightarrow (\text{cold water + agitation}) \rightarrow 4H_3PO_4$$

The reaction of *cold* phosphorus with atmospheric oxygen produces light, hence the term *phosphorescence* (from the Greek for "I bear" and "light"; see figure 11.12).

PHOSPHORUS PESTICIDES AND OTHER POISONS

Organophosphate esters and *phosphorothionates* are often extremely toxic compounds that have been used as pesticides and insecticides. The organophosphate pesticides are generally biodegradable and thus less likely to build up in soils and water. However, they are extremely toxic to humans, so great care must be used in handling and applying them. (See figure 11.13.)

- *Tetrethylpyrophosphate* is supertoxic (toxicity rating 6) to all mammals because it is a potent acetylcholinesterase inhibitor.
- *Phosphorothionates* have a greater selective toxicity for insects and readily biodegrade following metabolic conversion of $P = S$ to $P = O$.
- *Parathion* exists in liquid or solid form. It is also supertoxic because of its metabolic conversion to *paraoxon* (another potent acetylcholinesterase inhibitor).
- *Malathion* is a brown liquid with a skunk-like odor that can be metabolized to relatively nontoxic products by mammals, but not by insects, so its LD_{50} in male rats is approximately 100 times higher than that of parathion.

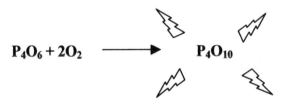

Figure 11.12. Phosphorescence reaction.

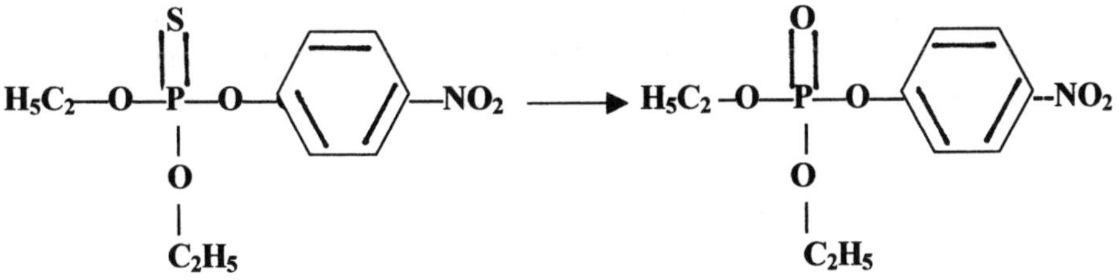

$$H_5C_2-O-\overset{\overset{\displaystyle S}{\|}}{\underset{\underset{\displaystyle C_2H_5}{|}}{P}}-O-\langle\bigcirc\rangle-NO_2 \longrightarrow H_5C_2-O-\overset{\overset{\displaystyle O}{\|}}{\underset{\underset{\displaystyle C_2H_5}{|}}{P}}-O-\langle\bigcirc\rangle-NO_2$$

Parathion   **Paraoxon**

Figure 11.13. Phosphorus pesticides.

Phosphorus poisons: Nerve gas military poisons, such as Sarin and VX, are lethal at extremely low doses (perhaps as low as 0.01 mg/kg; see figure 11.14).

PHOSPHOLIPIDS AND NUCLEIC ACIDS

Phospholipids are one of a group of lipids (fats) having both a phosphate group and one or more fatty acids. Although lipids are insoluble in water, the corresponding phospholipid has a water-soluble component. Therefore, these molecules can interact with both fatty materials and water-soluble materials, making them essential components of cell membranes (see figure 11.15).

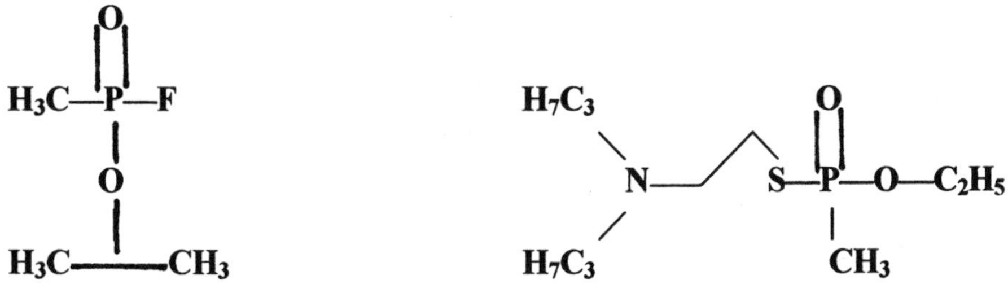

$$H_3C-\overset{\overset{\displaystyle O}{\|}}{\underset{\underset{\displaystyle H_3C-\!\!-\!\!-CH_3}{|}}{\underset{O}{P}}}-F$$

$$\underset{H_7C_3}{\overset{H_7C_3}{>}}N-\!\!\!\diagup\!\!\!\diagdown\!\!\!-S-\overset{\overset{\displaystyle O}{\|}}{\underset{\underset{\displaystyle CH_3}{|}}{P}}-O-C_2H_5$$

Sarin   **VX**

Figure 11.14. Phosphorus poisons.

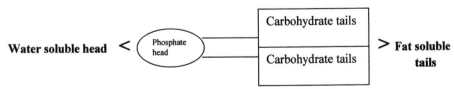

Figure 11.15. Essential components of cell membranes.

Phosphorus plays an important role in the structure of nucleic acids. Recall that the primary structure of both DNA and RNA consists of an alternating sugar phosphate backbone linked to nitrogen heterocyclic compounds called bases.

Sulfur Chemistry

Because sulfur (Sanskrit, *sulvere*; Latin, *sulphurium*; referred to in Genesis as brimstone) occurs free in nature as a solid, it has been known from very early times.

✔ **Key Point:** Sulfur is relatively abundant as an element in a solid state because it does not combine readily with other elements; it is the least reactive nonmetal.

In the periodic table, sulfur is found in Group VIA, has 6 outer-shell electrons and has a general tendency to acquire 2 addition electrons (oxidation state -2). However, other positive states are common, as in the sulfates (oxidation state $+6$).

✔ **Key Point:** Sulfur exhibits distinctly nonmetallic behavior. It oxidizes metals, giving a variety of binary sulfides in which sulfur exhibits a negative oxidation number.

Common naturally occurring sulfur compounds include:

- Sulfide ores (pyrite — FeS_2, Cu_2S, galena — PbS, As_2S_2, Sb_2S_3)
- Sulfate ores (gypsum — $CaSO_4$, $SrSO_4$, $BaSO_4$)
- Biologically significant sulfur species

Free sulfur is mined from enormous underground deposits in Texas and Louisiana. It is also obtained, in large quantities, from hydrogen sulfide recovered during the purification of "sour" natural gas.

Although the sources, fate, and transport of sulfur compounds in our atmosphere are not completely understood, the majority are emitted by way of human activities. Other sources include volcanic activity and the decay of organic matter.

Sulfur is used as a starting material to produce many commercially important materials, including sulfuric acid, matches, gunpowder, insecticides, fertilizers, paper, and medicines.

Important sulfur compounds are:

- Acids, esters, and salts
- Oxides
- Thiols/mercaptans
- Halides
- Pesticides and poisons
- Medicines
- Biologically significant compounds

SULFUR ACIDS, ESTERS, AND SALTS

Sulfuric acid (H_2SO_4), also known as *oleum* (fuming sulfuric acid) or *oil of vitriol*, is prepared by oxidizing sulfur to sulfur trioxide and then converting the trioxide to sulfuric acid. Pure sulfuric acid is a colorless, oily liquid that freezes at 10.5°C. It fumes when heated because the acid decomposes to water and sulfur trioxide. It is used to make nitric acid, hydrochloric acid, phosphate fertilizers, explosives, dyes, drugs, pigments, and food flavorings. It is also used in petroleum refining, electroplating, and storage batteries. Sulfuric acid is arguably the most significant chemical compound in industry.

Sulfurous acid (H_2SO_3) is unstable, and anhydrous sulfurous acid cannot be isolated. However, it combines readily with many coloring agents to produce colorless products, hence its use as a milder bleaching agent than chlorinated bleaches. However, because of its instability, the resultant colorless products gradually yellow with age.

The salts and esters of sulfuric acid are called *sulfates* (S = +6). See figure 11.16.

Sulfonates are esters in which C is attached directly to S. One of the first commercial detergents was a highly branched alkylbenzenesulfonate that was not biodegradable (see figure 11.17).

Manufacturers now produce biodegradable straight-chain alkylbenzenesulfonate and alkyl sulfate detergents.

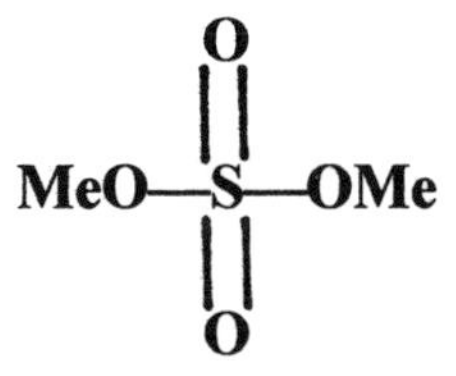

**Dimethyl sulfate
(ester)**

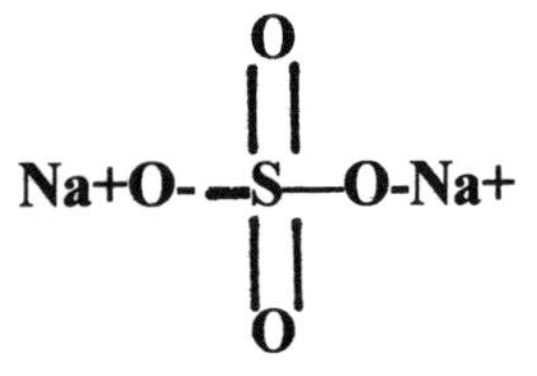

**Sodium sulfate
(salt)**

Figure 11.16.   Sulfates.

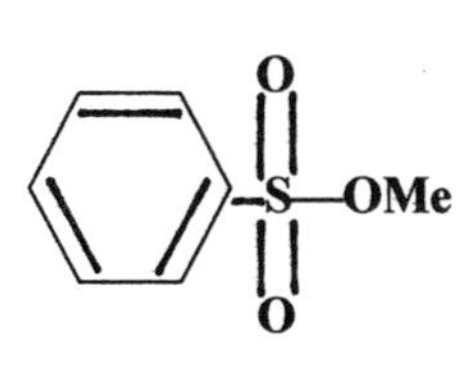

Methyl benzenesulfonate

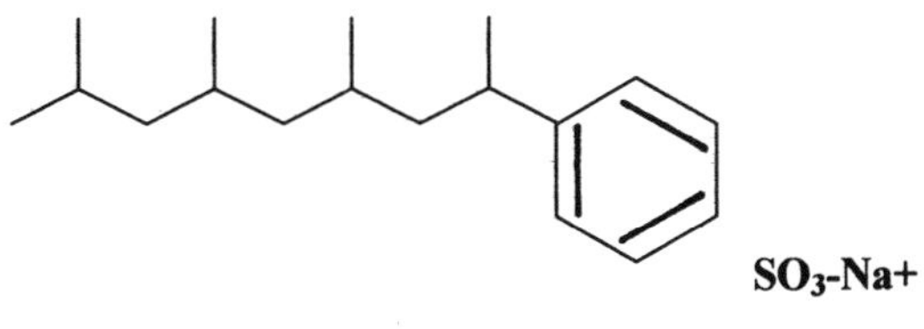

A branched chain sulfonate detergent

Figure 11.17.   Sulfonates.

The salts of sulfurous acid are called *sulfites* (S $=$ $+4$). Commercially important examples include Na_2SO_3 (*sodium sulfite*) and $Na_2S_2O_5$ (*sodium pyrosulfite*). Solutions of these sulfites, known commercially as *bisulfites*, are used in paper manufacture. Sulfites such as sodium sulfite, and sodium and potassium metabisulfite, were banned as food additives to raw fruits and vegetables in 1986. Sulfites still appear as food additives in other products, including wine, canned foods, pickles, etc. Numerous mediations also contain sulfite preservatives and stabilizers.

The artificial sweetener *cyclamate* (banned in 1970) is the sodium salt of a sulfamic acid (see figure 11.18).

OXIDES OF SULFUR

The two common oxides of sulfur are *sulfur dioxide* (SO_2) and *sulfur trioxide* (SO_3). The odor of burning sulfur is due to sulfur dioxide. It is used as a starting material for the manufacture of sulfuric acid. It is also used as a bleaching agent and preservatives for fruits, as well as a disinfectant in breweries and food manufacture. The National Ambient Air Quality Standard for SO_2 is 0.03 ppm (annual arithmetic mean) and 0.14 ppm (24-hour average).

When sulfur-containing coal is burned, it releases SO_2 to the atmosphere, leading to many detrimental effects, including leaf necrosis, respiratory distress, and acid rain.

$$4FeS_2(s) + 11O_2(g) \rightarrow 2Fe_2O_3(s) + 8SO_2(g)$$

A percentage of sulfur dioxide is also converted to sulfur trioxide (SO_3) in the atmosphere and in catalytic converters. SO_3 in the presence of water is converted to H_2SO_4.

$$SO_2(g) + 1/2O_2(g) \rightarrow SO_3(g)$$

Dimethyl *sulfoxide* (DMSO) is an important commercial solvent that has also been used to promote dermal absorption of medications (see figure 11.19).

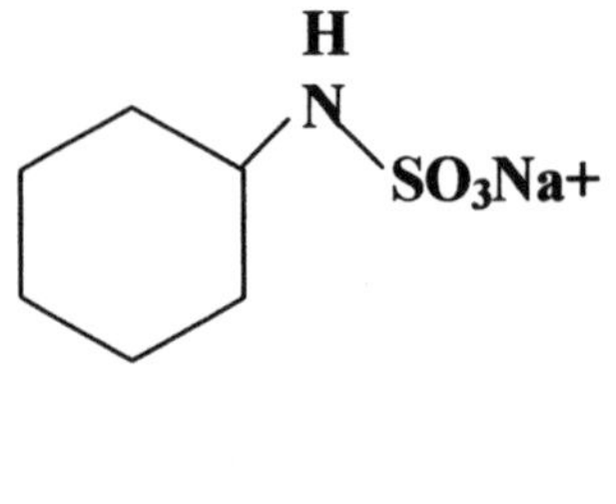

Sodium cyclohexylsulfamate

Figure 11.18. The sodium salt of a sulfamic acid (cyclamate).

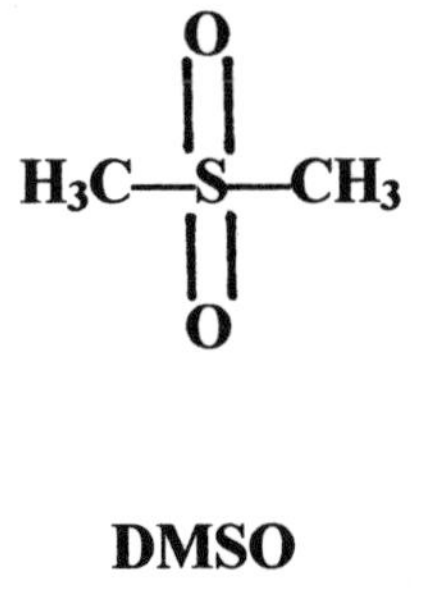

$$\text{H}_3\text{C}\!-\!\overset{\displaystyle \overset{\text{O}}{\|\|}}{\underset{\displaystyle \underset{\text{O}}{\|\|}}{\text{S}}}\!-\!\text{CH}_3$$

DMSO

Figure 11.19.   Dimethyl sulfoxide.

THIOLS/MERCAPTANS

The sulfur analog to an alcohol in which the oxygen atom is replaced by a sulfur atom is a *thiol* (formerly known as *mercaptans*). A characteristic property is their strong disagreeable odor. We detect these smelly compounds at concentrations as low as 0.02 ppb!

   Methanethiol (gas) and *ethanethiol* (volatile liquid) are added to natural gas, propane, and butane as leak detectors.

✔ **Key Point:** Other odor-producing organosulfur pollutants are commonly released into the atmosphere from packing houses, rendering plants, wood pulping, sewage treatment, and petroleum refining.

SULFUR HALIDES

Sulfur monochloride (S_2Cl_2) is used to vulcanize rubber, thus increasing its durability and temperature stability.

SULFUR PESTICIDES AND POISONS

Some of the best known sulfur-containing pesticides have already been discussed. Deadly *mustard oil* is a military poison which rapidly penetrates deep into tissue, causing "blistering" lesions that frequently become infected. Lesions in the lungs can cause death (see figure 11.20).

SULFUR IN MEDICINES

Some of the first truly effective antibacterial agents were the *sulfonamides* (sulfa drugs), which eventually gave rise to an entire family of bactericides (see figure 11.21).

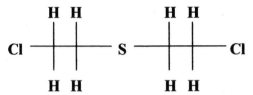

Figure 11.20.   Mustard oil.

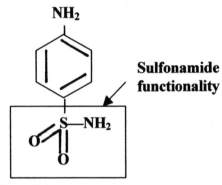

Figure 11.21.   Sulfur in medicines.

BIOLOGICALLY SIGNIFICANT SULFUR COMPOUNDS

The S-containing amino acid *cysteine* is often critical in determining the 3D shape of proteins. When 2 cysteines move close together through the natural folding of the long amino acid chains that make up proteins, a *sulfur* (disulfide link) can form, creating a covalent bond (see figure 11.22).

Pesticide Plan

There are more than 30,000 commercial pesticides on the market today, providing ample opportunity for exposure. Hayes points out that pesticides occupy a rather

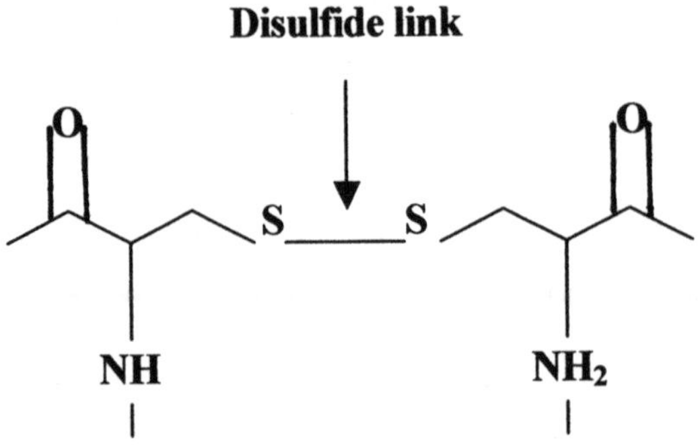

Disulfide link

Figure 11.22.   A sulfur disulfide link.

unique position among the many chemicals that humans encounter daily, in that they are deliberately added to the environment for the purpose of killing or injuring some form of life. Ideally their injurious action would be highly specific for undesirable target organisms and noninjurious to desirable, nontarget organisms. However, this is not always the case.[3]

The following is a list of various occupations exposed to pesticides.[4] Because of the large amount of pesticides in existence it is impossible to discuss details for each one; I will therefore only discuss the environmental practitioner's responsibilities in regard to pesticide control in the workplace.

- Treating contaminated workers
- Livestock dippers and veterinarians
- Research chemistry
- Agricultural application
- Structural application
- Flaggers
- Entomologists
- Workers on highway or railway rights-of-way
- Wood treatment workers
- Formulating end product
- Manufacturing active ingredient
- Mixing and loading
- Emergency work
- Sewer work
- Hazardous waste workers
- Medical personnel
- Vector control workers
- Park workers
- Building maintenance work
- Greenhouse, nursery, mushroom housework
- Firefighters
- Storage and warehouse work
- Farm work
- Crop duster maintenance

From the previous list, it should be obvious that the environmental practitioner may have his/her hands full in managing and controlling use of pesticides in the workplace.

In my experience, the average environmental professional knows that pesticides are harmful, that they require accompanying MSDSs, and that some training for proper application of the product is required. Beyond that, many practicing environmental professionals don't have a clue on how to manage and/or control pesticide usage in the workplace.

Beyond those requirements just mentioned, there are other concerns that must be taken into consideration in the proper management and control of pesticide usage in the workplace. For example, consider the following questions:

- Are workplace pesticide applicators state-certified or properly trained?
- Is a health monitoring protocol in place for pesticide-handling personnel; is such a protocol conducted or documented?
- Are pesticide-application records maintained at the plant?
- Are pesticide storage facilities not segregated from other non-related spaces?
- Are concentrated pesticides improperly released to sanitary wastewater or stormwater drains?
- Are banned pesticides used?
- Are pesticides stored in a dry, well-ventilated area designated for two air changes per hour?
- Are PPE, such as respirators, masks, gloves, coveralls, and equipment, available?
- Are restrictions on application of "restricted use" pesticides followed?
- Is there a *written* pesticide plan in place?
- Is the *written* pesticide plan administered, followed, and abided by?

While all the items listed above are important in controlling pesticide use in the workplace, it is the last two items that are the focus of our attention here.

Simply, environmental practitioners cannot be expected to perform their pesticide management/control functions properly unless there is a written plan.

SAMPLE PESTICIDE PLAN

Note: The following is a general pesticide plan that can be adapted to any EH&S operations involving pesticide use. The Appendixes A through D referred to in the pesticide plan will vary with the user's particular operations and are not supplied. The user's company name should, of course, appear at the initial reference to "the Company."

Purpose

One of the major goals of the Company's Occupational Health and Safety (OH&S) Department is to promote safe work practices in an effort to minimize the incidence of illness and injury experience by employees. In the USEPA's Federal Insecticide, Fungicide, and Rodenticide Act (FIFRA), 7 U.S.C §§136 et seq. (1972), the requirements governing the use pesticides were increased. The Company has instituted a Pesticide Plan [hereafter referred to as "the Plan") to ensure that pesticides are controlled. The Plan must:

- Identify pesticides used
- Maintain current MDSDs for the pesticides used
- Establish written safe work practices when required

General Program Management

There are three main categories of responsibility central to the effective implementation of the Plan.

- Occupational Health and Safety
- Department Heads and Managers
- Designated employees

The Company's OH&S department is responsible for the overall management and support of the the Company's compliance to OSHA regulations. Such responsibilities include, but are not limited to:

- Overall responsibility for implementing the Plan
- Working with management and employees to develop and administer policies and practices needed to support the effective implementation of the Plan
- Developing improvements to the Plan, as well as revising and updating the Plan where necessary
- Knowing current legal requirements
- Facilitating periodic on-site inspections to maintain an up-to-date Plan
- Planning and conducting all operations in accordance with Safe Work Procedure controls outlined in the Plan

Availability of the Exposure Control Plan to Employees

Copies of the Plan are kept in [employee accessible areas—lunchrooms, bulletin boards, etc.]. OH&S have additional copies available.

Review and Update of the Plan

To ensure the Plan is kept current, it will be reviewed and updated under the following circumstances:

- Annually, on or before 1 February each year
- Whenever new or modified tasks and procedures are implemented that affect occupational exposures of employees
- Whenever employees' jobs are so revised that new instances of occupational exposure may occur
- Whenever new positions are established in Company that may involve increased exposure

Definitions

Closed system—a device and procedure for transferring a pesticide from one container to another in a manner that does not expose the operator to the pesticide.

Field—any area, excluding a body of water, on or in which one or more crops are grown and includes but is not limited to a field of row crops, an orchard, a mushroom farm, a greenhouse, a nursery, a turf farm, a silviculture area and any similar area.

Fumigant—a pesticide applied in the form of a gas or vapor to kill pests and which is typically applied in an enclosed space.

Material Safety Data Sheet (MSDS)—a document produced by chemical manufacturers that describes physical and chemical properties of chemicals, first aid information, personal protective equipment (PPE) to use, and suggested means of disposal.

Pest—an injurious, noxious, or troublesome insect, fungus, weed, rodent, parasite, or other organism.

Pesticide—a microorganism or material that is represented, sold, used or intended to be used to prevent, destroy, repel, or mitigate a pest and includes: (1) a plant growth regulator, plant defoliator, or plant desiccant, and (2) a control product, other than a device, under the State/Local Pest Control Products Regulation [hereafter "Regulation"]

Restricted entry interval—the length of time representing a period of precaution that must elapse after the application of a pesticide, before an unprotected worker may be authorized to enter the treated portion of a building, structure, or field to which the pesticide has been applied.

Toxicity level—refers to the dangerous character of a pesticide, expressed as the Lethal Dose 50% (LD_{50}) by oral or dermal routes of entry of one containing active ingredients that have acute mammalian toxicities determined by an authority acceptable to OSHA/EPA; the lowest LD_{50} by the oral or dermal entry route determines the category of the pesticide and if the LD_{50} is reported as a range, the lowest reported LD_{50} is used.

Methods of Compliance

The Company enforces the following regulatory requirements:

- Pesticide labels
- Signage for treated areas
- MSDS availability
- Worker information
- Qualifications of applicator
- Procedures
- Rescue from area sprayed
- Restricted entry intervals
- Authorization to enter sprayed area

Pesticide Labels

The pesticide used has to be registered and labeled by the manufacture in accordance with pertinent regulations. The pesticide has to be used in accordance with the instructions on the label and follow good application practices.

Signage for Treated Areas

Signage has to include the following information if the pesticide leaves residues harmful to workers on plants for spraying:

- The name of the pesticide applied
- The hazards associated with its use
- The hazards and precautions required for handling these materials

MSDS Availability

MSDSs for the pesticides used at the Company must be available to workers at all times. See Appendix A [not included here] for a list of pesticides used at the Company. Appendix B [not included here] contains the MSDSs for those pesticides and is available in designated areas.

Worker Information

Having well informed and educated employees is very important when attempting to eliminate or minimize exposure to hazards. All employees who have the potential for exposure to a pesticide due to their proximity to the area where the pesticide is to be sprayed must be informed of:

- The intent to use the pesticide
- The hazards associated with its use
- The precautions required during the operation

Qualifications of Applicator

Any employee who mixes, loads, or applies a moderately or very toxic pesticide for use in the workplace or who cleans or maintains equipment used in pesticide applications has to be older than 18 and hold a valid applicator certificate issued under pertinent State/Local regulations.

Procedures

Many of the administrative controls have been in effect at the Company for some time. New safe work procedures are always being worked on. For the procedures relevant to pesticides review the Safe Work Procedures SOP. Extra copies can be obtained from OH&S. The *written safe work procedures* in Appendix E [not included here] are required for:

- Storage, handling, mixing, and application of pesticides
- Cleanup and disposal of spilled pesticides
- Summoning first aid and medical assistance for workers overexposed to pesticides

Rescue

When a moderately or very toxic pesticide is being applied in the greenhouse and the worker may be incapacitated during the application, the work must be done in such a manner that another worker equipped and able to do so can effect a rescue.

Equipment

Any equipment used to mix, load, or apply a pesticide must be:

- Constructed of materials chemically compatible with the pesticide being used
- Operated only by trained and authorized employees
- Used in accordance with instructions from the equipment manufacturer and pesticide supplier
- Maintained in a safe operating condition
- Cleaned, repaired, and maintained by workers who have been instructed in the necessary safe work procedures
- In a safe condition before maintenance or repair work is carried out

Warning Signs

Before a toxic pesticide or fumigant is applied the Company must ensure that:

- acceptable warning signs are conspicuously displayed at normal entrance points to the work area being treated; and
- if the pesticide is applied in an enclosed area, all entrances must be secured to prevent unauthorized access.

The warning signs for toxic pesticides, other than fumigants in enclosed spaces, must display:

- A skull and crossbones symbol
- The word *warning* in a language or languages that can be readily understood by workers in letters large enough to be read at a distance of 25 feet
- The name of the pesticide and the date of application
- The expiration date of the restricted entry interval
- Instructions to obtain permission to enter before the expiration date of the restricted entry interval

The warning sign for the application of a fumigant in an enclosed space must display:

- A skull and crossbones symbol
- The words DANGER, DEADLY FUMIGANT GAS, KEEP OUT in a language or languages that can be readily understood by workers and in letters large enough to be read at a distance of 25 feet

- The name of the fumigant
- The name of the applicator
- Night and day emergency telephone numbers

Entry to Treated Area

Personnel are not allowed to enter a treated area until the restricted entry interval is over. The length of the restricted entry interval is a minimum of:

- 24 hours for a slightly toxic pesticide
- 48 hours for a moderately or very toxic pesticide and for any mixture in which moderately or very toxic pesticides are present
- The interval specified on the pesticide label if that interval is longer than the above intervals

If, before the restricted entry period expires, the Company authorizes a worker to enter a field, building, or structure has been applied the Company must ensure that:

- the hazards to the worker have been assessed by a qualified person;
- the worker is provided with and is wearing the proper personal protective equipment (PPE); and
- the worker follows the proper procedures.

If the Company authorizes a worker to enter a building or structure in which a pesticide has been applied, the Company must ensure that, where practicable, the treated area is ventilated and the atmosphere has been tested or otherwise evaluated by a qualified person and declared safe to enter. If a worker may be incapacitated after reentry, provision must be made for rescue.

Personal Protective Equipment (PPE)

Personal protective equipment (PPE) must be used when a worker cleans, handles, or maintains equipment, materials, or surfaces contaminated with pesticide residues, the Company must ensure that:

- the worker is provided with and wears suitable protective clothing and equipment;
- contaminated protective clothing and equipment is stored in a secured place and is not used until it is laundered or otherwise cleaned;
- if needed, adequate facilities or services are available to launder contaminated protective clothing;
- there is at least one change of outer clothing for each worker; and
- a change room or sheltered place is provided where applicators can change clothes and store personal belongings while wearing protective clothing.

This equipment includes, but is not limited to:

- Chemical-resistant gloves
- Respirator and suitable cartridges
- Face shield/mask
- Safety glasses
- Chemical-resistant goggles
- Tyvek suit

Department heads are responsible for ensuring that appropriate PPE is available to employees. Training can be requested from OH&S on any type of PPE.

Employees are trained regarding the use of the appropriate PPE. Additional training is provided, when necessary, if an employee takes a new position or new job functions are added to their current position.

To ensure that PPE is not contaminated and is in the appropriate condition to protect employees from potential exposure, department heads must adhere to the following practices:

- All PPE is inspected periodically and repaired or replaced as needed to maintain its effectiveness.
- Reusable PPE is cleaned, laundered, and decontaminated as needed.
- Single-use PPE (or equipment that cannot be decontaminated) is disposed of according to the Company's Procedure for the Disposal of Hazardous Waste.

Storage

The pesticide storage shed has to meet the criteria listed in *State/Local Standard Practices for Pesticide Application*. Factors that must be considered include:

- Maintenance of minimal quantities
- Compatibility of pesticides
- Strength of the shelving materials
- Spill containment

Records

The Company maintains management of EH&S records and these are submitted to the OH&S office annually. Information provided includes:

- Name of pesticide used
- Rate of application
- Location
- Size of area treated
- Amount of product used
- Application method

If it is an external application, the windspeed and direction, temperature, and precipitation are also noted.

To comply with the Company regulation, the dates on which workers are allowed to reenter the area are maintained. A copy of the form is found in Appendix C [not included here].

Personal Hygiene

The Company has shower facilities in all locations for people involved in:

- mixing, loading, or applying pesticides;
- cleaning, maintaining, or handling equipment, materials, or surfaces contaminated with pesticide residues; and/or
- entering areas where pesticides have been applied and where contact with pesticide residues may contaminate protective clothing and body areas.

New Employee

When a new employee is hired, or an employee changes jobs within the Company, the employee must have a current pesticide applicators certificate to ensure that they are trained in the appropriate Safe Work Practice.

Pesticide Spills

Before a spill occurs a plan of how to deal with it must be in place. Adequate equipment must be available at locations where pesticides are stored and mixed and should include:

- Personal protective equipment
- Absorbent material
- Neutralizing agent
- Long-handled brush
- Shovel
- Waste-receiving container with lid

Procedure

Keep other people away from the spill. If the spill occurs on a roadway, prevent vehicles from traveling over the spilled material.

Review information about the pesticide about how to clean up the spill (MSDS, label)—this must be done before you start to use the pesticide, Once a spill happens, it is too late. Before starting cleanup, put on the correct personal protective equipment. If the spill is inside an enclosed area, ventilate the area. If the pesticide is flammable and explosive levels may be present, ensure the ventilation is explosion proof.

During cleanup—do not wash away the spilled pesticide as this only spreads it. Use the *BAN* system:

Barricade or dike the spilled pesticide to prevent its spread.

Absorb or soak up as much liquid material as possible. Absorbents include clay, vermiculite, and absorbent. Flammable absorbents such as rags, sawdust, and paper are less favorable than nonflammable absorbents. If the pesticide is a dust, wet down before sweeping. Dispose of the absorbent carefully—place in sealed, watertight drums labeled with the name of the pesticide. Then contact the OH&S office at [telephone number] for information about proper disposal.

Neutralize any remaining residue. Use a long-handled brush to scrub the spill area to help minimize inhalation or vapors. Specific neutralization techniques for the pesticides used at the Company are listed in Appendix D [not included here].

If you splash or spill a pesticide while mixing or loading, stop immediately. Remove contaminated clothing and wash thoroughly with detergent and water before reuse. Speed is essential when you or your clothing is contaminated.

Fire

Smoke from pesticide fires usually contains hazardous levels of unburned pesticide. All pesticide fires produce acid gases that can irritate the lungs. Some gases such as hydrogen sulfide and hydrogen cyanide are very toxic to life. Many organophosphates can be converted in fires to more toxic chemicals. A number of pesticides in containers can explode at higher temperatures.

Company has notified the location of the pesticide storage shed to the Regional Fire Department.

- If a fire occurs, evacuate personnel and animals that are downwind of the fire and keep people away.
- Call the switchboard and make it clear that it is a pesticide fire. Ask them to contact Company Emergency Response Team in addition to the Fire Department.
- When the Fire Department arrives, give them the pesticide inventory in Appendixes A and B [not included here].

Additional Reading

Carson, R. L. *Silent Spring*. Boston: Houghton Mifflin Company, 1993.
Lapierre, D., and J. Moro, trans. K. Spink. *Five Past Midnight in Bhopal*. Warner Books, Inc., 2002.
Matolesy, G., et al. *Pesticide Chemistry*. Elsevier Science, 1989.

Summary

- A *pesticide* is any substance or mixture of substances intended for preventing, destroying, repelling, or mitigating any pest.

- A *pest* is anything living that interferes with the smooth flow of life.
- Effective pesticide control in the workplace requires not only knowledge of pesticides, but also requires employee training.
- Nitrogen makes up about 78% of the atmosphere by volume.
- Apart from the atmosphere, nitrogen naturally occurs in soils due to the decomposition of animal matter by bacteria, manures, and plant and animal proteins.
- Nitrogen fixation can result in a wide variety of compounds because nitrogen compounds have a large range of oxidation states.
- Ammonia is produced in nature when any nitrogen-containing organic material decomposes in the absence of air.
- Alkaloids are naturally occurring amines.
- Approximately 75% of nitric acid is synthesized for the production of fertilizers.
- Ammonium nitrate is a well-known fertilizer that also has caused significant damage to lakes and streams due to eutrophication.
- Carbamate pesticides are common pesticides used in agriculture, as insecticides, fungicides, herbicides, nematocides, or sprout inhibitors.
- The basis for understanding biochemical genetics requires understanding of the chemical nature of nucleic acids.
- The primary structure of both DNA and RNA consists of an alternating sugar phosphate backbone linked to nitrogen heterocyclic compounds called bases.
- Pure, elemental phosphorus is rare.
- Phosphorus is commonly found in nature in the form of phosphates.
- Phosphate detergents have been implicated in eutrophication.
- Sulfur is the least reactive nonmetal and therefore large quantities are found free in nature.
- A pesticide plan is a tool for ensuring that pesticides are controlled.

New Word Review

Alkaloids—naturally occurring amines.

Ammonium nitrate—a well-known fertilizer.

Carbamate pesticides—common pesticides used in agriculture.

Closed system—a device and procedure for transferring a pesticide from one container to another in a manner that does not expose the operator to the pesticide.

Eutrophication—the natural aging of lakes.

Field—any area, excluding a body of water, on or in which one or more crops are grown and includes but is not limited to a field of row crops, an orchard, a mushroom farm, a greenhouse, a nursery, a turf farm, a silviculture, and any similar area.

Fumigant—a pesticide applied in the form of a gas or vapor to kill pests and which is typically applied in an enclosed space

Herbicide—weed killers; pesticides under FIFRA and are regulated as such.

Nitrogen—a gas that makes up 78% of the atmosphere by volume.

Nucleic acids—DNA and RNA, which store and pass on genetic information that controls reproduction and protein synthesis.

Pesticide—any substance or mixture of substances intended for defoliating or destroying, repelling, or mitigating any insects, rodents, fungi, bacteria, weeds, or other forms of plant or animal life or viruses, except viruses on or in living man or other animals.

Phosphorus (P)—a nonmetallic element.

Sulfur (S)—the least reactive nonmetal.

Chapter Review Questions

11.1.   A _____ is any substance or mixture of substances intended for preventing, destroying, repelling, or mitigating any pest.

11.2.   Drugs used to control diseases of humans or animals are not considered _____.

11.3.   Makes up 78% of the atmosphere: _____

11.4.   _____ are made up of amino acids.

11.5.   _____ is produced in nature when any nitrogen-containing organic material decomposes in the absence of air.

11.6.   _____ are naturally occurring amines.

11.7.   _____ is a well-known fertilizer that also has caused significant damage to lakes and streams.

11.8.   _____ is a highly toxic herbicide that rapidly disrupts enzyme and organ function.

11.9.   Excess nitrogen from protein metabolism is excreted in _____ and _____ as guanidine.

11.10. "Bone ash": _____

11.11. _____ is the least reactive nonmetal and therefore large quantities are found free in nature.

11.12. Also known as oleum: _____

11.13. Two common oxides of sulfur are _____ and _____.

11.14. A characteristic property is their strong disagreeable odor: _____

11.15. Regulates use of pesticides in U.S.: _____

11.16. A pesticide applied in the form of a gas or vapor to kill pests and which is typically applied in an enclosed space: _____

11.17. The acronym BAN stands for _____.

Notes

1. N. C. Brady and R. R. Weil, *The Nature and Properties of Soils*, 11th ed. (Upper Saddle River, N.J.: Prentice Hall, 1996), 604.

2. D. Lowenthal, "Awareness of Human Impacts: Changing Attitudes and Emphases," *The Earth as Transformed by Human Action* (Cambridge University Press, 1990), 121–35.

3. W. J. Hayes Jr., "Pesticides and Human Toxicity." *Ann. N. Y. Acad. Sci.* 160 (1969): 40–54.

4. R. Rosentack and M. R. Cullen, *Textbook of Clinical Occupational and Environmental Medicine* (Philadelphia: Harcourt Brace/W. B. Saunders, 1994).

CHAPTER 12

Toxicological Chemistry

All substances are poisons; there is none which is not a poison. The right
dose differentiates a poison and a remedy.

—Paracelsus (1493–1541)

Topics in This Chapter

- What Is Toxicology?
- Toxic Chemicals
- Risk Assessment
- Toxicological Chemistry and the Regulators

Toxicology focuses on the factors in our environment that have the potential to cause
adverse health effects. These factors may occur naturally or as a result of human
activities. Understanding the nature of these factors and their potential actions on
employees contributes to environmental practitioners making more informed decisions in environmental matters.

What Is Toxicology?

Toxicology has been defined as the study of the effects of chemical agents on biological
material, with special emphasis on harmful effects. There is ample evidence to indicate
that every chemical is capable, under some conditions, of producing some type of
effect on every biological tissue. Accordingly, you can simply define toxicology as the
science of poisons.

A poison, or *toxicant,* is a substance that is harmful to living organisms. Many
poisons are *xenobiotic materials* (i.e., all of the compounds that are foreign to the
organism; it also can be defined to include naturally present compounds administered
by alternate routes or at unnatural concentrations). Poisonous effects depend on:

- The type of organism exposed
- The amount of the substance
- The rate of the exposure by inhalation, injection, or skin contact

Physical forms of toxicants include:

- Gases
- Vapors
- Dusts
- Fumes
- Mists

Exposure depends on:

- Dose
- Toxicant concentration
- Duration
- Frequency
- Rate

Toxicologists try to identify:

- The *relationship* between the presence of a chemical and associated symptoms
- The *mechanisms* by which toxicants are transformed into other species
- The *processes* by which organisms eliminate toxicants and their metabolites
- *Antidotes*

Toxicological chemistry deals with the chemical nature of toxic substances, including their origins, uses, fate, and transport.

EXPOSURE TO TOXIC CHEMICALS

Toxicologists focus on (1) the toxicity of the same substance to different organisms; (2) the toxicity of different organisms; and (3) minimum doses for toxic effects. Other important variables include:

- Sensitivity to small increases in dose
- Determination of the lethal dose
- Reversibility of toxic effects
- Acute/intermediate/chronic effects

✔ **Key Point:** *Acute* toxicity and *intermediate* toxicity refer to responses observed soon after exposure to a toxic substance.

✔ **Key Point:** *Chronic toxicity* refers to effects that manifest themselves long after exposure.

Toxicants that leave no permanent effect, through natural defense mechanisms or antidotes, are said to be *reversible*.

Dose is defined as the degree of exposure of an organism to a toxicant. The observed effect to a dose is the *response*. *Dose-response curves* demonstrate the relationship between the dose and the observed effect (see figure 12.1).

Response to a very low level of toxicant is called *hyper*sensitivity. Response to only very high levels of toxicant is referred to as *hypo*sensitivity.

Substances may be assigned toxicity ratings, ranging from nontoxic (>15 mg/kg body mass) to super toxic (<5 mg/kg body mass).

TOXICANT METABOLISM

Toxicants in the body can be absorbed, metabolized, seek temporary storage, or be distributed or excreted. Stated differently, toxicants may pass through the body unchanged, be detoxified, be converted to a toxic active metabolite, have adverse biochemical effects, and/or cause manifestations of poisoning. *Immediate manifestations of poisoning* include changes in vital signs:

- Temperature
- Pulse rate
- Respiratory rate
- Blood pressure

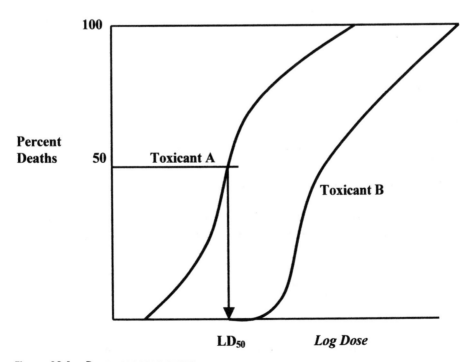

Figure 12.1. Dose response curve.

Other evidence of poisoning includes:

- Odors
- Changes in eyes
- Convulsions
- Paralysis
- Hallucinations
- Coma

Still other toxic responses may include:

- Impairment of enzyme function
- Alteration of cell membranes
- Interference with carbohydrate and lipid metabolism
- Interference with respiration
- Interference with protein biosynthesis
- Interference with regulatory processes

Acute and intermediate responses may also occur, such as:

- Gastrointestinal illness
- Cardiovascular symptoms
- Liver and kidney disease
- Neurological symptoms
- Skin abnormalities

Chronic responses include:

- Mutations
- Cancer
- Birth defects
- Immune system effects

IMPORTANT TERMINOLOGY

- The *immune system* is the body's natural defense mechanism against xenobiotic chemicals, infectious agents, and neoplastic cells (which give rise to cancerous tissue).
- Toxicants can suppress the immune system (*immunosupression*).
- Toxicants can also cause *allergic reactions* and *hypersensitivity*.
- *Chemical carcinogenesis* refers to the role of foreign substances in causing uncontrolled cell replication (i.e., cancer).

✔ **Key Point:** Frequently carcinogens form covalent bonds with macromolecules such as DNA, allowing cells to reproduce uncontrollably.

- *Primary carcinogens* cause cancer directly.
- *Precarcinogens or procarcinogens* require metabolic activation by various mechanisms to produce *ultimate carcinogens*.
- The *Bruce Ames test* determines carcinogenicity based on mutagenicity. Mutagens trigger the reversion of histidine-dependent Salmonella bacteria back to histidine-synthesizing bacteria.
- *Mutagens* alter DNA to produce inheritable traits and can often also cause birth defects.
- *Teratogens* are chemicals that cause birth defects, typically through damage to embryonic or fetal cells.
- *Synergistic effects* are well-documented adverse interactions between certain substances that make them far more hazardous in combination than individually. A synergistic effect occurs when the observed effects are greater than what would be predicted by simply summing the effects of each individual chemical. For example, chemical X produces 30% mortality, and chemical Y produces 30% mortality. However, if the effects are synergistic, when chemicals X and Y are administered together the mortality is greater than 60%. As an example, both carbon tetrachloride and ethanol are toxic to the liver, but together they produce much more liver injury than the sum of their individual effects. Other synergistic effects are common:
  ○ Smokers are $10 \times$ more likely to develop lung cancer than nonsmokers.
  ○ Asbestos workers are at a significantly higher risk of developing lung cancer.
  ○ An asbestos worker who smokes is $90 \times$ more likely to develop lung cancer than a person exposed to neither cigarette smoke nor asbestos.

    Smoking + alcohol consumption
    = higher risks of cancer of the mouth, larynx, and esophagus
    Smoking + living in highly polluted areas
    = higher risks of lung cancer
    Smoking + working with chemicals
    = higher rates of lung cancer[1]

- *Selective toxicity* is used to eliminate parasites from the human body. Penicillin is toxic to a number of microorganisms but is relatively innocuous to humans. This technique can be used because of the differences in metabolism and cell structure of the host organism and the parasite.
- Occupational health organizations use toxicological data and other information to produce health guidelines for chemical hazards in the workplace. The key organizations include:
  ○ OSHA—Occupational Safety and Health Administration
  ○ NIOSH—National Institute for Occupational Safety and Health
  ○ ACGIH—American Conference of Governmental Industrial Hygienists
In 1974, NIOSH joined OSHA in developing a series of complete occupational standards. Source for these guidelines included recognized references in toxicology, analytical chemistry, occupational medicine, and industrial hygiene. Among other data, these guidelines provide:

- ○ Exposure limits
- ○ IDLH values
- ○ Recommendations for PPE
- ○ Route(s) of Health Hazard
- ○ Symptoms
- ○ First aid
- ○ Target organs
- Exposure Limits; Threshold Limit Values (TLVs): The ACGIH publishes a booklet—*Threshold Limit Values*—updated every year, concerning exposure limits in the workplace and carcinogenic potential for a variety of chemicals. The exposure levels are referred to as Threshold Limit Values (TLVs). The TLVs ("refers to airborne concentrations of substances and represent conditions under which it is believed that nearly all workers may be repeatedly exposed day after day without adverse health effect") are used as guidelines in establishing the acceptable risk of exposure to a toxic substance. The TLVs are based on information obtained from human and animal studies as well as on experiences from various industries. "TLVs are based on the best available information from industrial experience, from experimental human and animal studies, and when possible, from a combination of the three."
- IDLH Values: The Immediately Dangerous to Life or Health (IDLH) value represents "a maximum concentration from which one could escape within 30 minutes without any escape-impairing symptoms or any irreversible health effects."[2]
- Recommendations for PPE/Sanitation: These guidelines provide suggestions for the following categories:
- ○ Clothing
- ○ Goggles
- ○ Washing
- ○ Changing (protective clothing)
- ○ Removing (protective clothing)
- ○ Providing eyewash/quick drench equipment
- Routes of Health Hazard: The toxicologically important routes of entry are also identified:
- ○ Inhalation
- ○ Skin absorption
- ○ Ingestion
- ○ Skin and/or eye contact
- Symptoms and First Aid: The potential symptoms of exposure as well as recommended first aid procedures are also provided.
- Target Organs: The organs that are affected by exposure are also identified.

Toxic Chemicals

The quantity of chemical, physical, and biological agents with the potential to cause toxic effects seems almost unlimited. In this text, I concentrate on specific toxic chemicals including ozone, heavy metals, silica, asbestos, and inorganic sulfur compounds.

OZONE (O_3)

Ozone has a distinct odor at 1 ppm and can cause severe irritation and headache, fatal pulmonary edema, chromosomal damage, and free radical destruction of lipids, sulfhydryl (-SH), and other important functional groups.

HEAVY METALS

Metals exist in rocks and soils that form Earth's crust. Humans use these metals for various purposes. My focus is on some of the more frequently encountered metals. Information on the effects of exposure to these metals is well documented in numerous sources.

Cadmium (Cd) is a soft, white-bluish metal, a by-product of the mining and smelting of lead- and zinc-containing ore. It is used in electroplating and galvanizing processes because it is resistant to corrosion. NIOSH reports that approximately 4,000 metric tons of cadmium are used yearly in the United States. About half of this is used for plating other metals, and the rest is used in pigments, batteries, stabilizers for plastics, metallurgy, nuclear reactor neutron-absorbing rods, and semiconductors and as a catalyst. Approximately 1.5 million workers may be potentially exposed to cadmium. Exposure to cadmium interferes with the function of certain key enzymes and can also cause bone disease, kidney damage, and pulmonary edema and pulmonary epithelial necrosis.

Lead (Pb) is one of the most widely used and broadly distributed toxic substances known. The primary use of lead is in the manufacture of lead-acid batteries, which accounts for almost 70% of all the lead consumed in the United States. Examples of jobs associated with lead exposure:

- Lead production or smelting
- Battery manufacturing
- Brass, copper, or lead foundries
- Radiator repair
- Demolition of old structures
- Construction
- Thermal stripping or sanding of old paint
- Scrap metal handling
- Indoor firing ranges
- Lead soldering
- Ceramic glaze mixing
- Welding of old painted metal
- Machining and grinding

Exposure to lead can cause inhibition of hemoglobin synthesis, neural disorders, and disruption of kidney function. *Tetraethyl lead* [$Pb(C_2H_5)$], which was widely used as a gasoline additive, affects the central nervous system, causing such symptoms as fatigue, weakness, psychosis, and convulsions.

Arsenic (As) is a metalloid and the twentieth most abundant element present in Earth's crust. The effects of exposure to arsenic depend on the chemical state of the compound as well as the length of exposure. Arsenic exposure has shown to coagulate proteins, complex important coenzymes, and inhibit key metabolic processes.

Mercury (Hg), like lead, has received much attention because of its toxic effects, which vary according to the element's physical state. At room temperature, elemental mercury is a silvery liquid. Liquid mercury volatizes at room temperature, releasing mercury vapor. Exposure to mercury in the workplace is common through breathing or skin contact in dental or health services and chemical and other industries that use mercury. The blood can carry mercury to the brain, where it causes tremors and neuropathological disorders such as insomnia, depression, and irritability.

Silica (SiO_3), quartz, contained in dust is generated from a variety of occupations. Fine silica-containing particles are airborne and inhaled. Examples of industries and work activities that pose the greatest potential risk for worker exposure include:

- Construction: sandblasting, rock drilling, masonry work, jack hammering, and tunneling
- Mining: cutting or drilling through sandstone and granite
- Foundry work: grinding and molding and working in shakeout and core room
- Stone cutting: sawing, abrasive blasting, clipping, and grinding
- Railroad: setting and laying track

Silica dust can cause silicosis, a lung disease that causes pulmonary fibrosis and increased susceptibility to other lung diseases.

Asbestos [$Mg_3P(Si_2O_5)(OH)_4$] has a number of unique properties. It has a high melting point; it is resistant to degradation when exposed to heat and chemicals, and it has low heat conductivity. Asbestos is composed primarily of silica, bound to various other elements such as magnesium, iron, and chromium, in the form of fibers. Asbestos ends up in commercial products like brake pads and roofing material intentionally. But it also ends up in consumer products by accident. For example, many lawn-care products contain vermiculite. Unfortunately, when that vermiculite is ored, it may contain traces of asbestos. As a result, the asbestos ends up in a bag of fertilizer—not on purpose—but through contamination. Inhalation of asbestos fibers can cause asbestosis (a pneumonia carcinoma, or cancer of the bronchial tubes).

INORGANIC SULFUR COMPOUNDS

Hydrogen sulfide (H_2S) is a colorless, flammable gas that is heavier than air. It has a distinct "rotten egg" odor, which may be detected at concentrations as low as 0.025 ppm. It is found in natural gas and oil, in mines, wells, fertilizers, sewers, and cesspools. It is given off as a by-product in the manufacture of rayon, synthetic rubber, dyes, and the tanning of leather. Hydrogen sulfide kills faster than HCN (hydrogen cyanide), with death occurring from respiratory paralysis at concentrations around

1,000 ppm. Lower doses result in headaches, dizziness, and central nervous system damage. It also saturates nasal receptors, "freezing" the olfactory sense, thereby becoming "odorless" (i.e., undetectable by smell) at higher concentrations.

Sodium sulfite (Na_2SO_3) was used as a food preservative until severely restricted in the United States in 1990 because of individual hypersensitivity to this compound.

Sulfuric acid (H_2SO_4) is produced in greater quantity than any other chemical in the United States. It is used in the production of a variety of other chemicals such as alcohol, ammonium sulfate, hydrofluoric acid, and aluminum sulfate. It is also used in the manufacturing of fertilizers, batteries, cellophane, and titanium dioxide. Sulfuric acid is a corrosive poison that causes severe tissue necrosis and dehydration, as well as eye and respiratory irritation.

TOXIC ORGANIC CHEMICALS

PCBs are halogenated aromatic hydrocarbons. PCBs may cause atrophy of the various lymphoid organs, such as the spleen, and reduce the number of circulating lymphocytes as well as antibody production.

Pesticides (organophosphate and carbamate pesticides) are widely used. Poisoning from exposure to organophosphates is common in rural areas and in developing countries. Exposure to these pesticides may occur by inhalation, ingestion, or dermal absorption. Symptoms are dependent on the route of exposure.

Dioxin is one of the most toxic chemicals known; burning chlorine-based chemical compounds with hydrocarbon forms it. The major source of dioxin in the environment (95%) comes from incinerators burning chlorinated wastes. A draft report released for public comment in 1994 by the USEPA clearly describes dioxin as a serious public health threat. The public health impact of dioxin may rival the impact that DDT had on public health in the 1960s. According to the USEPA report, not only does there appear to be no "safe" level of exposure to dioxin, but levels of dioxin and dioxin-like chemicals have been found in the general U.S. population that are "at or near levels associated with adverse health effects." The USEPA report confirmed that dioxin is a cancer hazard to people. Dioxin is very fat-soluble and is not easily transported to the brain, because of its affinity for plasma proteins.

✔ **Key Point:** One of the dangers of using pesticides such as DDT or chemicals such as dioxin is biomagnification, which causes an increase in concentrations of a contaminant (DDT or dioxin) from one link in a food chain to another. In order for biomagnification to occur, the toxin must be long-lived, mobile, soluble in fat, and biologically active.

✔ **Interesting Point:** The classic example of biomagnification is that of DDT. During the years that it was used in the United States, DDT caused thinning eggshells in many birds of prey, including eagles and hawks. The pesticide was present at low concentrations in water because of runoff from agricultural fields. But DDT is very persistent and accumulates in fats. The lower doses in lower trophic levels (feeding levels) resulted in no observable adverse effects. However, the high dose accumulated by fish-eating birds caused thinning of eggshells and reduced reproductive success, resulting in drastic declines in populations of these species nationwide.[3]

Risk Assessment

Risk, relative to toxic effects, is defined as the ratio of the number affected to the number exposed. For example, the lifetime risk at birth of developing any form of cancer is 23 cancers for every 100 U.S. citizens. USEPA's "acceptable risk" for developing cancer from a hazardous waste site is 1 cancer per 1 million exposed. Examples of 1/1,000,000 risk of death:

- Motor vehicle accident: 1.5 days
- Fire: 13 days
- Lightning: 2 years
- Animal bite or sting: 4 years

Examples of 1/1,000,000 cancer risk:

- 1/7 of a chest X-ray
- Smoking 2 cigarettes
- Camping at 15,000 feet for 6 days compared to sea level
- 2.5 months in masonry rather than wood structure

In regards to *acceptable risk,* W. B. Katz makes the point that "the decision of where to concentrate efforts to reduce a hazard raises the question of what constitutes an acceptable risk. This subject affects all of us all the time, and what we are willing to accept depends to a great extent on or perception of the risk, how much we know about it, and how much we think we can control it."[4]

As an example of acceptable risk, riding a motorcycle provides adventure and pleasure, at the risk of being thrown and hurt. With the large number of motorcycles seen every day on any interstate highway in the United States, it is obvious that many are willing to accept this risk.

Because there is no such thing as "zero risk," all of us must *manage* our risks to maintain our own level of comfort as individuals, families, corporations, and agencies. Often without even realizing it, we make important risk management decisions for ourselves and our organizations every day:

- Do I get into my car at rush hour?
- Do I have adequate knowledge to review and sign this hazardous waste manifest, or should I ask for additional assistance?
- Do I conduct additional health and safety training for my workers?
- Do I don a respirator before entering a chemical process area?
- Do I sample the air before entering a confined space?

Risk assessment is the process of gathering all available information on the toxic effects of a chemical and evaluating it to determine the possible risk associated with exposure. Risk assessments are a major tool for toxicologists to determine the estimated potential danger of pollutants.

Risk assessments help predict the potential effects of accidental exposures, environmental pollution, environmental cleanup, workplace exposure, and residual contamination levels at hazardous waste sites. Risk assessments are designed to:

- Characterize the "dose-response relationship"
- Numerically estimate the risk of the occurrence of a health effect
- Numerically estimate the number of cases expected
- Characterize the uncertainty of the analysis
- Recommend "acceptable" concentrations of chemicals in the air, food, water, and soil

BASIC RISK MODEL

Source
Release
    Transport Media
Transport
      Receptor
      Exposure
         Receptor Physiology
         Toxic Effect

Risk assessment begins with gathering and evaluating information. "The process of gathering and evaluating the information can be divided into four steps: 1) hazard identification, 2) hazard evaluation or dose-response assessment, 3) exposure assessment, and 4) risk characterization."[5]

Hazard Identification

Hazard identification consists of the collection of data (from various sources) that are necessary to determine whether a substance is toxic. It requires performance of toxicity testing to determine a toxicity range (in the absence of toxicity data). The adverse human health effects are also considered.

The following list (adapted from USEPA's *Risk Assessment Guidelines and Information Directory*) provides the various types of information that should be identified and examined when performing the hazard identification:

1. Identify substance (name)
2. Physical/chemical properties of the toxic substance
   A. Solubility, chemical reactivity, molecular size, ion/non-ion state

3. Source of toxicity information
   A. Epidemiological studies
      (1) prospective study
      (2) retrospective study
   B. Toxicological study
      (1) Acute toxicity studies: Thresholds, no-observed-effect-level (NOEL), no-observed-adverse-effect-level (NOAEL), LD_{50}, LC_{50}
      (2) Chronic toxicity study: Mutagenic, teratogenic, carcinogenic, other biological effects (heart failure, liver disease, skin rash, etc.)
      (3) Species of test animal
      (4) Other variables affecting toxicity: Age, sex, health
4. Exposure to toxic substance
   A. Route of exposure
      (1) Skin contact
      (2) Inhalation
      (3) Ingestion
      (4) Injection
   B. Duration of exposure
   C. Frequency of exposure
   D. Exposure to other toxic substances
5. Other confounding factors
   A. Diet, lifestyle choices, occupation

Hazard Evaluation

If the hazard identification process produces evidence of a hazard, then a hazard evaluation is performed. Tools used to complete this process include:

- Dose-response relationships
- Chemical potency
- Species variation
- Mechanism(s) resulting in adverse effect

Risk assessments help predict the potential effects of:

- Accidental exposures
- Environmental pollution
- Environmental cleanup
- Workplace exposure
- Residual contamination levels at hazardous waste sites

Risk assessments are designed to:

- Characterize the "dose-response relationship"
- Numerically estimate the risk of the occurrence of a health effect

- Numerically estimate the number of cases expected
- Characterize the uncertainty of analysis
- Recommend "acceptable" concentrations of chemicals in the air, food, water, and soil

Exposure Assessment (Routes of Exposure)

Exposure assessment is performed to identify the affected population, and, if possible, quantify the magnitude, frequency, duration, and route of exposure. The following list (adapted from USEPA's *Risk Assessment Guidelines and Information Directory*) provides the various factors that should be considered when performing an exposure assessment:

1. General information for each chemical
   A. Identity
       (1) Molecular formula and structure
       (2) Other identifying characteristics
   B. Chemical and Physical Properties
2. Sources
   A. Characterization of Production and Distribution
   B. Uses
   C. Disposal
   D. Summary of Environmental Releases
3. Exposure Pathways and Environmental Fate
   A. Transport and Transformation
   B. Identification of Principal
   C. Predicting Environmental Distribution
       4. Measured or Estimated Concentrations
   A. Uses of Measurements
   B. Estimation of Environmental Concentrations
5. Exposed Populations
   A. Human Populations
       (1) Population size and characteristics
       (2) Population location
       (3) Population habits
6. Integrated Exposure Analysis
   A. Calculation of Exposure
       (1) Identification of the exposed population
       (2) Identification of pathways of exposure

Risk Characterization

This is the final step in the risk assessment. Using the information gathered in the other steps, mathematical modeling of animal and/or human toxicity data, combined with human exposure evaluation, risk is characterized (estimated).

Estimation of the *risks of adverse health effects* typically involves extrapolation from experimentally observable data. Data are generally obtained from acute high doses of toxicants to animals, followed by extrapolation to chronic low doses of exposure in humans.

The Food and Drug Administration (FDA), Consumer Products Safety Commission (CPSC), USEPA, and OSHA base policy on risk assessment data. Many regulations and policies evolve from the concept of "acceptable risk" or "reasonable risk."

✔ **Key Point:** If adverse human effects are predicted by the risk assessment, then further analysis of potential ecological effects may be warranted. In terms of environmental risk management, be aware of your facility's risk management policy when making decisions that could impact the entire organization.

Epidemiology studies may also be performed to correlate observations of particular illnesses with probable exposure. They help to identify diseases known to be caused by particular agents. They also look for clusters of abnormally high occurrences of a particular disease (e.g., spontaneous abortions, birth defects, and cancer).

✔ **Key Point:** Epidemiology studies can be complicated by long latency periods, lack of specificity in correlation to specific diseases, and background levels of disease.

Toxicological Chemistry and the Regulators

For the toxicologist and/or the general environmental practitioner responsible for determining toxicity of various chemicals and chemical compounds used in the workplace, various U.S. regulatory agencies such as OSHA and EPA can provide valuable information.

The regulators' information developed in the 1970s and 1980s when the most significant increase in the passage of laws and the growth of government agencies occurred. The new laws addressed the manufacture, use, and disposal of chemicals.

OSHA

The Centers for Disease Control and Prevention (CDC) website[6] can provide the EH&S practitioner with the latest information about OSHA regulations for hazardous chemicals. The website is the source of the following information.

According to the OSHA *Hazard Communication Standard* (1910.1200), a hazard determination must consider the chemicals listed in the following sources to be hazardous:

- Chemicals regulated by OSHA in 29 CFR Part 1910, Subpart Z.
- *Threshold Limit Values for Chemical Substances and Physical Agents in the Work Environment*, American Conference of Governmental Industrial Hygienists (latest edition).
- National Toxicology Program, *Annual Report on Carcinogens* (latest edition).
- International Agency for Research on Cancer Monographs (latest edition).

✔ **Key Point:** The fact that a chemical is not listed does not mean it is not hazardous. Any chemical that presents a potential health or physical hazard to which employees may be exposed must be included in the hazard communication program.

USEPA

To help meet its objective USEPA has developed and promulgated regulations that implement the requirements of several comprehensive environmental laws. These laws and regulations set standards and requirements for the manufacturing, use, and disposal of thousands of toxic chemicals. The specific chemicals regulated by USEPA and the associated standards are found in the *Code of Federal Regulations*, or *CFR*. The CFR contains the implementing language and regulatory requirements for all federal laws that are passed.

One of the most important USEPA regulations affecting the environmental practitioner in the workplace is the *Toxic Substances Control Act* (TSCA) of 1976. TSCA requires testing and reporting of chemicals prior to manufacturing, distribution, and use and restricts the use of chemicals that pose a threat to human health and the environment.

TSCA also gives USEPA the ability to track 75,000 industrial chemicals currently produced or imported into the United States. USEPA repeatedly screens these chemicals and can require reporting or testing of those that may pose an environmental or human-health hazard. USEPA can ban the manufacture and import of those chemicals that pose an unreasonable risk.

TSCA requires USEPA to classify substances as either "existing" chemicals or "new" chemicals. The only way to determine if the substance you are working with is a new chemical is by consulting USEPA's *Toxic Substances Control Act Chemical Inventory*. Any substance that is not on the inventory is classified as a new chemical.

Moreover, any substance not on the list that is flammable, explosive, toxic, or corrosive is also considered to be hazardous and is reportable if discharged.

Additional Reading

Manahan, S. E. *Toxicological Chemistry*. Boca Raton, Fla.: CRC Press, 1992.
Stelljes, M. E. *Toxicology for Non-Toxicologists*. Rockville, Md.: Government Institutes, 2000.
USEPA. *Risk Assessment Guidelines and Information Directory*. Rockville, Md.: Government Institutes, 1988.

Summary

- Toxicology focuses on the factors in our environment that have the potential to cause adverse health effects.
- Toxicology has been defined as the study of the effects of chemical agents on biological material with special emphasis on harmful effects.
- A poison, or toxicant, is a substance that is harmful to living organisms.
- Poisonous effects depend on the type of organism exposed, the amount of the substance, and the rate of the exposure by inhalation, injection, or skin contact.
- Toxicological chemistry deals with the chemical nature of toxic substances, including their origins, uses, fate, and transport.
- Dose is defined as the degree of exposure of an organism to a toxicant.
- Dose-response curves demonstrate the relationship between the dose and the observed effect.
- Toxicants in the body are metabolized, transported, and excreted.
- The quantity of chemical, physical, and biological agents with the potential to cause toxic effects seems almost unlimited.
- Risk is defined as the ratio of the number affected to the number exposed.
- Risk assessment is the process of gathering all available information on the toxic effects of a chemical and evaluating it to determine the possible risk associated with exposure.
- Hazard identification consists of the collection of data that are necessary to determine whether a substance is toxic.
- Hazard evaluation is performed if the hazard identification process produces evidence of a hazard.
- Exposure assessment is performed to identify the affected population, and, if possible, quantify the magnitude, frequency, duration, and route of exposure.
- Risk characterization is the final step in the risk assessment.
- Estimation of the risks of adverse health effects typically involves extrapolation from experimentally observable data.
- According to the OSHA *Hazard Communication Standard*, a hazard determination must consider the chemicals listed on various authoritative lists.
- The Toxic Substances Control Act (TSCA) requires testing and reporting of chemicals prior to manufacturing, distribution, and use.

New Word Review

Acceptable risk—a risk that is so low that no significant potential for toxicity exists, or a risk society considers is outweighed by benefits.

Carcinogen—a cancer-causing substance.

Dose-response—the relationship between the amount of a chemical taken into the body and the degree of toxic response.

Extrapolation—using data from direct observations, typically laboratory animal tests, to predict results for unobserved conditions.

Metabolism—transformation of a chemical within an organism to other chemicals through reactions.

Mutagens—chemicals that alter DNA to produce inheritable traits and can often also cause birth defects.

Reversible effect—an effect that dissipates with time following cessation of chemical exposure.

Risk—the probability of an adverse effect resulting from an activity or from chemical exposure under specific conditions.

Risk assessment—a scientific process used to estimate possible exposures, cancer risks, and noncancer adverse health effects from known or estimated levels of chemicals.

Synergistic effects—observed effects that are greater than what would be predicted by simply summing the effects of each individual chemical.

Target organ—the primary organ where a chemical causes noncancer effects.

Teratogen—a chemical causing a mutation in the DNA of a developing offspring

Toxicity—the intrinsic degree to which a chemical causes adverse effects.

TSCA—U.S. Toxic Substances Control Act.

Unacceptable Risk—a risk that is perceived to be high enough to represent a significant potential for toxicity, or a risk society considers is not outweighed by benefits.

Xenobiotic—a chemical foreign to a living organism.

Chapter Review Questions

12.1.   What type of risk is it when we chose to ignore warnings about sticking to a healthy diet? _____

12.2.   _____ focuses on the factors in our environment that have the potential to cause adverse health effects.

12.3.   Poisonous effects depend on: _____.

12.4.   _____ deals with the chemical nature of toxic substances, including their origins, uses, fate, and transport.

12.5.   _____ is defined as the degree of exposure of an organism to a toxicant.

12.6.   Response to only very high levels of toxicant is referred to as _____.

12.7.   _____ carcinogens cause cancer directly.

12.8.   _____ is used to eliminate parasites from the human body.

12.9.   _____ are based on the best available information from industrial experience, from experimental human and animal studies, and when possible, from a combination of the three.

12.10.   _____ are the organs that are affected by exposure are also identified.

12.11.   _____ is one of the most toxic chemicals known.

12.12.   _____ is defined as the ratio of the number effected to the number exposed.

12.13.   _____ is the process of gathering all available information on toxic effects of a chemical and evaluating it to determine the possible risk associated with exposure.

12.14.   May be performed to correlate observations of particular illnesses with probable exposure: _____.

12.15. The following requires testing and reporting of chemicals prior to manufacturing, distribution, and use: _____.

Notes

1. A. Nadakavukaren, *Our Global Environment: A Health Prospective*, 5th ed. (Prospect Heights, Ill.: Waveland Press, 2000).

2. K. R. Olson, *Poisoning and Drug Overdose* (Appleton & Lange, 1994).

3. "Bioconcentration, Bioaccumulation and Biomagnification," *Ecorisk Fundamentals*, http://newweb.ead.anl.gov/ecorisk/fundamentals/html (accessed December 30, 2002).

4. W. B. Katz, *The ABCs of Environmental Science* (Rockville, Md.: Government Institutes, 1998), 91.

5. C. Kent, *Basics of Toxicology*. (New York: John Wiley & Sons, 1988).

6. Centers for Disease Control and Prevention, *OSHA Designated Hazardous Chemicals*, http://www.cdc.gov/od/ohs/manual/chemical/chmsaf7.htm (accessed December 30, 2002).

CHAPTER 13

Nuclear Chemistry and Radioactivity

In 1991 the widow of a construction worker who helped build the British Nuclear Fuels (BNF) Sellafield plant was awarded a large sum of money when it was determined that his death from chronic myeloid leukemia was the result of overexposure to radiation. Sellafield was constructed for the purpose of separating uranium from used fuel rods. Working at the plant for approximately nine months, the victim received a total cumulative dose of almost 52 millisieverts of radiation, which exceeded the established limit for an entire twelve-month period. BNF compensated the victim's wife and the families of twenty additional workers who died from causes related to radiation.[1]

Radiation hazards in many workplaces are a real and present danger to workers. This chapter provides prospective and practicing environmental health and safety professionals with the basic information they need concerning nuclear chemistry.

Topics in This Chapter

- Nuclear Chemistry: The People
- Types of Radiation
- Penetration of Matter
- Alpha and Beta Radiation
- Fission and Fusion
- Half-Life
- Radiocarbon Dating
- Mixed Waste

Nuclear Chemistry: The People

E. O. Lawrence invented the cyclotron, which was used at the University of California at Berkeley to make many of the transuranic elements. Transuranic elements are those having atomic numbers greater than 92 (i.e., having more protons than uranium).

Lawrence's invention played an important role in the early days of nuclear radiation experiments. However, there were other scientists who preceded the work of Lawrence and are credited with making some of the most important discoveries concerning nuclear radiation. William Roentgen (1845–1923) was the first major player. He discovered X-rays, a high-energy form of light (1895). A year later, in 1896, Henri Becquerel (1852–1908) found that uranium ores emit radiation that can pass through objects (like X-rays) and affect photographic plates. Marie Curie (1867–1934) and Pierre Curie worked with Becquerel to understand radioactivity. The three shared the Nobel Prize in physics in 1903. Marie won a second Nobel Prize in chemistry in 1911 for her work with radium and its properties.

Ernest Rutherford (1871–1937), one of the first and most important researchers in nuclear physics, was very interested in the Curies' new discoveries. He obtained a small sample of radium and conducted his own experiments. Rutherford discovered that the ray actually consisted of three rays: a positive ray, a negative ray, and an uncharged ray. The positive ray (alpha particle) was found to consist of helium nuclei (a unit of two protons and two neutrons). The negative ray (beta particle) revealed itself to be a beam of electrons. The uncharged ray (gamma particle) was high-energy X-ray. Rutherford's studies of radiation led to his formulation of a theory of atomic structure, which was the first to describe the atom as a dense nucleus about which electrons circulate in orbit.

Types of Radiation

Radioactivity refers to the particles that are emitted from nuclei as a result of nuclear instability. As mentioned, there are three types of radioactivity:

Alpha particles (α) contain two protons and two neutrons. The alpha particle is a nucleus of the element of helium. Because of its very large mass (more than 7,000 times the mass of the beta particle) and its charge, it has a very short range.

Beta particles (β) consists of a stream of electrons (from the decomposition of a neutron into a proton and an electron).

Gamma rays (γ) are radiant energy similar to X-rays.

Alpha, beta, and gamma radiation is referred to as *ionizing radiation* because each kind dislodges electrons from the atoms and molecules it encounters.

✔ **Important Point:** Ionizing radiation can be very damaging to living tissue, causing cancers, birth defects, mutations, and death.

Penetration of Matter

Though the most massive and most energetic of radioactive emissions, the alpha particle is the shortest in range because of its strong interaction with matter. A sheet of ordinary paper can stop the alpha particles. It is the least dangerous. The electron of beta radioactivity strongly interacts with matter and has a short range. A layer of clothing can stop the beta particle. It is more penetrating and therefore more dangerous than alpha. The electromagnetic gamma ray is extremely penetrating, even penetrating considerable thickness of concrete. It takes several feet of concrete or several inches of lead to stop it. It can cause severe damage.

✔ **Key Point:** Nuclear chemistry involves only the *nucleus* (protons and neutrons), *not* electrons.

Alpha and Beta Radiation

Uranium nucleus (^{92}U) "decays" into a thorium atom (^{90}Th) by the loss of an *alpha particle*.

$$^{92}U \rightarrow alpha \rightarrow {}^{90}Th$$

A radium nucleus (^{88}Ra) becomes a radon nucleus (^{86}Rn) by the same process.

$$^{88}Ra \rightarrow alpha \rightarrow {}^{86}Rn$$

✔ **Important Point:** Radon is part of the natural radioactive decay chain of uranium.

Radon is a *Naturally Occurring Radioactive Material*. NORM refers to materials not covered under the International Atomic Energy Agency whose radioactivity has been enhanced usually by mineral extraction or processing activities. Examples are exploration and production wastes from the oil and natural gas industry and phosphate slag piles from the phosphate mining industry.[2] Exposure to NORM is often increased by human activities, such as burning coal and using fertilizers. Radon in homes (particularly those built on granitic ground) is one occurrence of NORM that may need to be controlled, by ventilation. Radon dilutes to harmless levels in the open air, but it can accumulate to hazardous levels in confined spaces (e.g., basements); hence, the need for ventilation. High levels of radon are dangerous because it decays to harmful alpha emitters such as polonium-218 and 214. When inhaled, they can be deposited in the lungs and increase the risk of lung cancer.

✔ **Key Point:** USEPA has estimated that radon, a common NORM, causes 7,000 to 30,000 additional lung cancer deaths annually, making it the *second leading cause of lung cancer deaths* in the United States.

Beta particles are negatively charged high-speed electrons and have much less mass than alpha particles. As mentioned, they are more penetrating and therefore more dangerous than alpha.

A neptunium nucleus (^{93}Np) is transformed into a plutonium nucleus (^{94}Pu) by the loss of a beta particle.

$$^{93}Np \rightarrow beta \rightarrow {}^{94}Pu$$

Gamma rays are extremely high-energy light, with no mass, are the most penetrating form of ionizing radiation, and *generally accompany beta radiation.* They can cause severe damage. *X-rays* are slightly less energetic but have generally the same characteristics as gamma rays.

✔ **Key Point:** Ordinary chemical reactions involve only electrons, but in nuclear reactions, the nuclei of atoms change, causing a *transmutation* of one element into another.

Fission and Fusion

Fission is the breaking apart of a nucleus into smaller and/or more stable nuclei. *Fusion* is the combining of nuclei into larger nuclei. Both processes release energy, but both require extremely high activation energies. Scientists have harnessed fission to produce electricity in nuclear power plants. Nuclear power plants produce energy (heat) by the controlled fission of uranium-235. Fission of U-235 produces a *chain reaction.*

Nuclear fission becomes self-sustaining when the number of neutrons produced by fission equals or exceeds the number of neutrons absorbed by splitting nuclei plus the number lost to the surroundings. The heat produced is used to make steam, which drives generators.

$$^{235}U + N \rightarrow Kr\text{-}90 + Ba\text{-}142 + 3n + heat$$

Figure 13.1 depicts a fission chain reaction. Fission chain reactions release an enormous amount of heat and energy within seconds. They must be controlled to prevent an explosion, such as the atomic bomb.

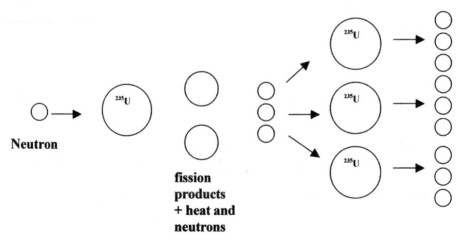

Neutron

fission products + heat and neutrons

Figure 13.1. A fission chain reaction.

Table 13.1. The half-life of various elements.

Element	Half-life
Uranium-238	4.51×10^9 years
Uranium-237	6.7 years
Uranium-239	23.5 minutes
Radium-226	1622 years
Radium-222	38 seconds

Scientists have not yet been able to harness the enormous power of nuclear *fusion*, but the sun fuses hydrogen atoms into helium to pour energy out into space.

$$4H \rightarrow heat \rightarrow He$$

Half-Life

Because radioactive atoms decay, they are said to be unstable. Chemists determine the stability of a nucleus by measuring its half-life. The half-life of a radioactive element is the time required for half of a radioactive sample to decay. Decay means that atoms of a particular isotope spontaneously change into new atoms. The shorter time for the half-life to occur, the more unstable the nucleus of the atom (the proton to neutron ratio is more unstable). Table 13.1 lists the half-life of various radioactive elements.

Half-life in chemistry is different from half of anything in math. Mathematically, if you keep walking half the distance to the store, you will never reach the store. The number scale contains an infinite set of possible numerical fractions. However, in chemistry the number of atoms is limited (usually around a mole—6.02×10^{23} of atoms). Despite the large number of atoms in a mole, it would take only about 70 half-lives to have an entire mole of atoms decay.

In basic chemistry, the question we most often ask is "What fraction of a radioactive isotope exists after (2, 3, 4, . . .) half-lives?" This can be determined using exponents shown in table 13.2. The stability of a nucleus, and hence its tendency to emit radioactive materials, depends on the balance among

Table 13.2. Half-Life Exponents.

No. of Half-Lives	Fraction Remaining
0	$(1/2)^0 = 1 = 100\%$
1	$(1/2)^1 = 1/2 = 50\%$
2	$(1/2)^2 = 1/4 = 25\%$
3	$(1/2)^3 = 1/8 = 12.5\%$
4	$(1/2)^4 = 1/16 = 6.25\%$
5	$(1/2)^5 = 1/32 = 3.125\%$
6	$(1/2)^6 = 1/64 = 1.5625\%$
7	$(1/2)^7 = 1/128 = 0.78125\%$

- attractive gravitational forces;
- repulsive electronic forces; and
- the ratio of protons to neutrons.

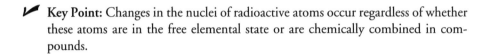 **Key Point:** Changes in the nuclei of radioactive atoms occur regardless of whether these atoms are in the free elemental state or are chemically combined in compounds.

Radiocarbon Dating

Radiocarbon dating determines the age of organic matter such as ancient mummies. The procedure was discovered in 1947 by a chemist, Willard Libby. Small amounts of carbon dioxide (CO_2) containing carbon-14 are produced in the upper atmosphere by cosmic rays (high-energy subatomic particles from space). Because the amount of radioactive CO_2 remains relatively constant, this radioactive CO_2 becomes equally distributed throughout all living plant and animal tissue. Thus, all living organisms absorb radiocarbon, an unstable form of carbon (C-14) that has a half-life of about 5,570 years. By measuring the amount of the radioactive isotope, such as carbon-14, that is left in the organism, an approximate age of an organism can be determined.

Mixed Waste

USEPA's website,[3] on which the material in this chapter is based, provides a basic understanding of the regulations that govern mixed waste.

The fundamental and most comprehensive statutory definition of mixed waste is found in the Federal Facilities Compliance Act (FFCA) where Section 1004(41) was added to RCRA: "The term *mixed waste* means waste that contains both hazardous waste and source, special nuclear, or by-product material subject to the Atomic Energy Act (AEA) of 1954."

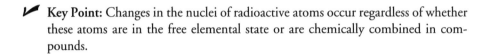 **Key Point:** Once a waste is determined to be a mixed waste, the waste handlers must comply with both AEA and RCRA statutes and regulations.

Almost all of the commercially generated (non-DOE) mixed waste is composed of low-level radioactive waste and hazard waste and is called Low Level Mixed Waste (LLMW). Commercially generated LLMW is produced in all fifty states at industrial, hospital, and nuclear power plant facilities.

Under the 1984 amendments to RCRA, Land Disposal Restriction (LDR) regulations prohibit disposal of most mixed waste, including LLMW, until it meets specific treatment standards.

Additional Reading

Saland, L. *Contemporary Chemistry: A Practical Approach* (Portland, Me.: J. Weston Walch, Publisher, 1991).

Summary

- E. O. Lawrence, William Roentgen, Henri Becquerel, Marie and Pierre Curie, and Ernest Rutherford were the early major players in studying nuclear radiation and making important discoveries.
- Radioactivity refers to the particles that are emitted from nuclei as a result of nuclear instability.
- There are three types of radioactivity: alpha particles, beta particles, and gamma rays.
- Alpha rays have little penetrating ability and can be stopped by a piece of paper.
- Beta rays are more penetrating and therefore more dangerous than alpha.
- Gamma rays are high-energy light and are the most penetrating form of ionizing radiation.
- Radon is a Naturally Occurring Radioactive Material (NORM).
- Fission is the breaking apart of a nucleus into smaller and/or more stable nuclei.
- Fusion is the combining of nuclei into larger nuclei.
- Radiocarbon dating determines the age of organic matter such as ancient mummies.
- Mixed waste is waste that contains hazardous waste and source, special nuclear, or by-product material subject to the Atomic Energy Act.

New Word Review

Atomic fission—the splitting of heavy atomic nuclei into lighter nuclei, with the release of much energy.

Atomic fusion—the merging together of light atomic nuclei into heavier nuclei, with a great release of energy.

Chain reaction—a self-sustaining radioactive decay process.

Electron—a light charged particle "revolving" around the nucleus of an atom.

Half-life—the rate of decay of a radioactive element, specific to each element, that measures the time for half of the element to be used up.

Neutron—a heavy particle in the nucleus of an atom that does not carry an electrical charge.

Proton—the heavy charged mass in the nucleus of an atom.

Radioactive decay—the spontaneous breakdown of radioactive elements into other elements or isotopes, with the emission of one or more types of radiation.

Radioactivity—the property of an element that causes it to spontaneously decay into an isotope of the same element or another element entirely, giving off radiation in the process.

Radon—a radioactive gas formed from the decay of uranium.

Transmutation—the change of one element into another.

Chapter Review Questions

13.1.   He discovered X-rays: _____

13.2.   _____ contain two protons and two neutrons.

13.3.   _____ consist of a stream of electrons.

13.4.   _____ are similar to X-rays.

13.5.   Radioactive particle with shortest range: _____

13.6.   _____ is part of the natural radioactive decay chain of uranium.

13.7.   _____ refers to materials whose radioactivity has been enhanced by mineral extraction or processing activities.

13.8.   _____ is the breaking apart of a nucleus into smaller and/or more stable nuclei.

13.9.   _____ is the combining of nuclei into larger nuclei.

13.10. The stability of a nucleus is determined by measuring its _____.

13.11. Contains hazardous waste and source, special nuclear, or by-product material subject to the AEA: _____

13.12. By measuring the amount of the _____, such as carbon-14, that is left in the organism an approximate age of an organism can be determined.

13.13. The half-life of a substance is 12.4 hours. How much of a 750-gram sample is left after 62.0 hours? _____

13.14. A 50-gram sample of a substance decays to 13.5 grams in 14.4 seconds. What is its half-life? _____

Notes

1. D. L. Goetsch, *Occupational Safety and Health,* 2nd ed. (Englewood Cliffs, N.J.: Prentice-Hall, 1996), 293.

2. USEPA, *Mixed Waste Glossary,* www.epa.gov/radiation/mixed-waste/mw_pg5.htm (accessed January 2, 2003).

3. USEPA, *About Mixed Waste,* www.epa.gov/radiation/mixed-waste/mw_pg3.htm.

CHAPTER 14

Hazardous Waste and Treatment Methods

In regard to hazardous waste, what must we do? "What we must do can be compared to what an efficient wastewater treatment plant does. When we produce waste, we must treat it so that the end product is sent back to its natural environment cleaner than it was in the first place. We do it constantly and consistently in wastewater treatment. The question is, can't we do it with all the wastes we produce? We think that goal is possible—especially if the scientific, political, social, and monetary commitment that needs to be made is made . . . and it will be. What other choice do we have?"[1]

Topics in This Chapter

- "Characteristic" Hazardous Waste
- "Listed" Hazardous Waste
- Exceptions to Hazardous Waste Designation
- Hazardous Waste Treatment Methods

Chemical spills can be the environmental professional's worst nightmare. For that matter, handling, storing, and disposing of hazardous waste is not exactly a walk in the park. Underpinning this view is the definition of what hazardous waste is.

USEPA defines hazardous waste as any waste or combination of wastes that pose a substantial present or potential hazard to human health or living organisms because such wastes are nondegradable or persistent in nature or because they can be biologically magnified, or because they can be lethal, or because they may otherwise cause or tend to cause detrimental cumulative effects.[2]

Environmental professionals deal with chemicals in the workplace on a consistent basis. Protecting employees from the detrimental effects posed by chemical exposure is a full-time job. MSDSs and other pertinent OSHA standards provide environmental professionals with guidelines to follow in protecting employees (and themselves) from potential harm from the chemicals they use in doing their work.

✔ **Important Point:** As essential as OSHA standards are in protecting workers from the chemicals they use, environmental professionals must to be aware that hazardous *waste* can be just as dangerous as duly packaged and labeled hazardous materials. It is also important to remember that when a hazardous material or chemical is spilled into the environment, it can no longer be categorized as such; it is now a hazardous *waste*. In dealing with hazardous wastes, both OSHA and USEPA guidelines play an important role in guiding those affected through mitigation procedures that are designed to protect not only the worker but also the environment.

One of USEPA's most important rules dealing with hazardous waste is the Resource Conservation and Recovery Act (RCRA). Under RCRA (Subtitle C), hazardous wastes are required to have "cradle-to-grave" tracking of their generation, transportation, treatment, storage, and disposal. RCRA lists two main types of hazardous waste: "characteristic" wastes and "listed" wastes.

"Characteristic" Hazardous Waste

"Characteristic" hazardous wastes are defined in the U.S. Code of Federal Regulations (40 CFR Part 261.20–24, Subpart C) as exhibiting one of four hazardous characteristics: ignitable, corrosive, reactive (or explosive), or toxic, but are not listed or otherwise excluded from Subtitle C regulation. These wastes are designated by the letter "D" followed by a three-digit code:

• Ignitability (D001)
• Corrosivity (D002)
• Reactivity (D003)
• Toxicity (D004–043)

✔ **Key Point:** It is the responsibility of the generator to determine whether his wastes exhibit one or more of these characteristics.

IGNITABILITY

An ignitable hazardous waste bears USEPA number D001. Any one of the following descriptions defines an ignitable condition in terms of the federal regulations:

• A liquid (other than aqueous solution containing less than 24% alcohol by volume) having a flash point of less than 140°F.
• Nonliquids that can catch fire by friction, absorption of moisture, or spontaneous chemical changes, and burn vigorously on catching fire.
• A compressed flammable gas as defined by the Department of Transportation (DOT) regulations (49 CFR 173.300).
• A waste defined as an oxidizer by DOT regulations (49 CFR 173.151).

✔ **Key Point:** To most easily determine if the substance qualifies as a "liquid," the Paint Filter Test (an SW846 Test Method) can be used. Flash point is determined by the American Society for Testing & Materials (ASTM) standard D-93-79 or D-93-80. *Aqueous* means that the material contains 50% or more of water by weight.

CORROSIVITY

A corrosive waste has USEPA number D002. Though some states have additional criteria, a waste is deemed to be corrosive if it exhibits one of the following conditions:

- It is an aqueous solution that has pH of 2 or less, or 12.5 or more.

or

- It is a liquid that corrodes steel at a rate of more than one-quarter inch (1/4) per year (SW846 Test Method 1110).

✔ **Key Point:** Recall that pH is a scale running from 1 to 14 (standard units), used to measure the strength of acidic and alkaline substances mixed with water or other liquids. Acidic solutions register below 7, and alkaline solutions register above 7, while the pH of pure water is 7, which is a neutral. The pH of solutions can also vary, depending on how dilute or concentrated they are. In addition, pH can be temperature sensitive, as solutions generally can be made more concentrated at higher temperatures. Vinegar generally has a pH of 3.4 to 2.4. The pH of a room-temperature solution of baking soda is about 8.3, the pH of household cleaning ammonia is approximately 11, and solutions containing lime or lye can be mixed to a maximum strength nearing pH 14. Some SW846 test methods are available.[3]

REACTIVITY

A reactive waste has USEPA number D003. These wastes are extremely unstable and have a tendency to react violently or explode in a number of situations, as listed by USEPA. A waste is deemed to be reactive if it exhibits one of the following conditions:

- It is normally unstable and readily undergoes violent changes without detonation (40 CFR Part 261.23(1)).
- It reacts violently with water (40 CFR Part 261.23(2)).
- It forms potentially explosive mixtures with water (40 CFR Part 261.23(3)).
- It generates dangerous quantities of toxic gases when mixed with water (40 CFR Part 261.23(4)).
- It is a cyanide- or sulfide-bearing waste that generates dangerous quantities of toxic gases when exposed to noncorrosive pH conditions (40 CFR Part 261.23(5)).

- It is capable of detonating or exploding when subjected to a strong initiating force or heated under confinement (40 CFR Part 261.23(6)).
- It is readily detonated or explosively decomposed at standard temperature and pressure (40 CFR Part 261.23(7)).
- It is a forbidden Class A or Class B explosive (DOT)—(40 CFR Part 261.23(8)).

✔ **Key Point:** Wastes characterized as reactive can be highly unstable during routine transportation, storage, and disposal. Few SW846 test methods are available.

TOXICITY

Toxic characteristic (TC) hazardous wastes are those solid wastes given USEPA numbers D004–D043, depending upon the toxic substance present. The TC is determined using a Toxicity Characteristic Leaching Procedure (TCLP). The TCLP tests for 25 organic chemicals, 8 inorganics, and 6 pesticides at or above specific federally regulated thresholds. That is, if the extracts leached from a waste using the TCLP test contain one or more of 39 toxic contaminants listed in Table I of 40 CFR 261.24 (see table 14.1)—and that table includes arsenic, benzene, cadmium, chlordane, cresol, lead, mercury, selenium, and vinyl chloride—at levels that exceed the maximum unregulated concentrations, the waste must be characterized as a toxic hazardous waste.

✔ **Key Point:** A hazardous waste is assigned *every* applicable EPA hazardous waste number, which must be used for all notifications, recordkeeping, and reporting requirements.

✔ **Interesting Point:** A solid waste is a characteristic hazardous waste if it meets one or more of the four hazardous waste characteristics.

"Listed" Hazardous Waste

The inclusive listing adopted by USEPA includes separate lists of nonspecific source wastes (*F List*), specific source wastes (*K List*), and commercial chemical products (*P List and U List*). USEPA developed these lists by examining different types of wastes and chemical products to determine whether they met any of the following criteria:

- Exhibit one or more of the four characteristics of a hazardous wastes
- Meet the statutory definition of hazardous waste
- Are actually toxic or acutely hazardous
- Are otherwise toxic

These lists, defined in the U.S. Code of Federal Regulations (40 CFR 261, Subpart D), include approximately 900 different hazardous, or acutely hazardous, wastes specifically "listed" in RCRA.

Table 14.1.   Toxicology characteristics.

EPA HW No.	Contaminant	Regulatory Level (mg/L)
D004	Arsenic	5.0
D005	Barium	100.0
D018	Benzene	0.5
D006	Cadmium	1.0
D019	Carbon tetrachloride	0.5
D020	Chlordane	0.03
D021	Chlorobenzene	100.0
D022	Chloroform	6.0
D007	Chromium	5.0
D024	m-Cresol	200.0
D023	o-Cresol	200.0
D025	p-Cresol	200.0
D026	Cresol	200.0
D016	2,4-D	10.0
D027	1,4-Dichlorobenzene	7.5
D028	1,2-Dichloroethane	0.5
D029	1,1-Dichloroethylene	0.7
D030	2,4-Dinitrotoluene	0.13
D012	Endrin	0.02
D031	Heptachlor (and its epoxide)	0.008
D032	Hexachlorobenzene	0.13
D033	Hexachlorobutadiene	0.5
D034	Hexachloroethane	0.3
D008	Lead	5.0
D013	Lindane	0.4
D009	Mercury	0.2
D014	Methoxychlor	10.0
D035	Methyl ethyl ketone	200.0
D036	Nitrobenzene	2.0
D037	Pentachlorophenol	100.0
D038	Pyridine	5.0
D010	Selenium	1.0
D014	Silver	5.0
D039	Tetrachloroethylene	0.7
D015	Toxaphene	0.5
D040	Trichloroethylene	0.5
D041	2,4,5-Trichlorophenol	400.0
D042	2,4,6-Trichlorophenol	2.0
D017	2,4,5-TP (Silvex)	1.0
D043	Vinyl chloride	0.2

The F List consists of 28 different waste streams from nonspecific sources. It includes some types of spent solvents, industrial wastewater treatment sludges, quench water sludges, petroleum-processing sludges, and leachate (liquids which have percolated through land-disposed wastes) from hazardous waste.

The K List, on the other hand, includes 116 hazardous wastes consisting mostly of residues from specific types of chemical reactions or distillation or purification processes associated with producing wood preservatives, organic and inorganic chemicals, pesticides, explosives, inks, veterinary pharmaceuticals, petroleum refining, metal refining, coke production, and coal tar distillation.

The P List includes 239 chemical substances that have been identified as acutely hazardous, although they also may be reactive or exhibit other characteristics. The P List consists of commercial chemicals that may be off-specification (for example, they may have impurities), might have been spilled, or are a container residue. If these P List substances are to be discarded, they must be disposed of as listed hazardous wastes. However, P List substances—whether off-specification, only partially spent, collected from nearly empty containers, or otherwise useful—may still be used in other commercial or industrial processes in lieu of disposal, and they are not to be considered as hazardous wastes until such time as they do require disposal.

The U List includes 521 substances identified as toxic wastes. They may have additional hazardous properties such as being ignitable, reactive, or corrosive. Like the chemical substances on the P List, U List substances also are commercial chemicals in need of disposal as a result of being spilled, being off-specifications, or because they are container residue. However, they too may be suitable for less stringent commercial or industrial process in lieu of disposal, and they are therefore not regulated as waste under such circumstances.

Exceptions to Hazardous Waste Designation

In addition to USEPA designations of "characteristic" and "listed" hazardous wastes, USEPA also designates Exceptions to Hazardous Waste Designation. Exceptions include:

- Carbon tetrachloride, tetrachloroethylene, and trichloroethylene discharged in small quantities (i.e., 1 ppm of total discharge) into a publicly owned treatment works (POTW) that is regulated under the Clean Water Act.
- Methylene chloride, 1,1,1-trichloroethane, chlorobenzene, 0-dichlorobenzene, cresols, cresylic acid, nitrobenzene, toluene, methyl ethyl ketone, carbon disulfide, isobutanol, pyridine, and spent chlorofluorocarbon solvents in small quantities (i.e., < 1 ppm of total discharge) into a POTW that is regulated under Section 307b or 402 of the Clean Water Act.
- Listed petroleum-refining by-products defined in Title 40 CFR Part 261.32.
- Insignificant losses from manufacturing.
- Wastewater from laboratory operations, provided that total wastewater flow is < 1% annualized and hazardous wastes are < 1 ppm of total flow.
- Used oil containing more than 1,000 ppm total halogens is assumed hazardous; however, if applicant can show that oil does not contain listed hazardous wastes, oil is assumed to be not hazardous.
- Household wastes.

- Mining overburden returned to the mine site.
- Fly ash waste, bottom ash waste, slag waste, and flue gas emission control waste generated from the combustion of fossil fuels.
- Drilling fluids and by-products of oil exploration.
- Chromium-contaminated wastes under certain circumstances (see 261.4(b)(6)(I)).
- Solid waste from the extraction, beneficiation (i.e., the dressing of some ores), and processing of some ores and minerals.
- Cement kiln dust.
- Some arsenical treated wood.
- Petroleum-contaminated media and debris if the corrective actions are regulated under Title 40 CFR Part 280 (Underground Storage Tank regulations).
- Used chlorofluorocarbons that are to be recycled.
- Some oil filters that are not terne-plated (i.e., not coated with a mixture of lead and tin).

Hazardous Waste Treatment Methods

Even with the vigorous hazardous waste reduction program that RCRA and Pollution Prevention (P²) programs mandate, large quantities of hazardous wastes will still require treatment (remediation) and ultimate disposal.

Treatment methods are often categorized as being chemical, physical, biological, thermal, or stabilization/fixation. Here, my focus is on the physical and chemical treatment methods only. Physical and chemical (along with biological) treatments are currently the most commonly used methods of treating aqueous hazardous waste.

Physical treatment methods include gravity separation, phase change systems, such as steam and air stripping of volatiles from liquid wastes, and various filtering operations, including carbon adsorption.

Chemical treatment transforms waste into less hazardous substances using such techniques as pH neutralization, oxidation or reduction, and precipitation.

Choosing an appropriate treatment method to use in any given situation is beyond the scope of this text. Not only are there many different kinds of hazardous wastes, in terms of their chemical makeup, but the treatability of the wastes depends on their form.[4]

PHYSICAL TREATMENT METHODS

Physical treatment methods are used to separate phases and components in a liquid waste stream. It primarily includes sedimentation, adsorption, and aeration.

Sedimentation

Basically, sedimentation is a phase separation method in which suspended particles settle under the influence of gravitational forces. Special sedimentation tanks are designed to encourage solids to settle so they can be collected as sludge from the bottom of the tank. Some solids will float to the surface, and they can be removed with a skimming device. Evaporation, filtration, or centrifugation can then further concentrate separated sludges.

Adsorption

Adsorption can be used to remove a wide variety of contaminants from liquid or gaseous streams. This process involves contacting the waste stream with a medium (e.g., granulated activated carbon—GAC) when the latter removes soluble organics and certain inorganics present in the wastewater. The effectiveness of the adsorbent is directly related to the amount of surface area available to attract the molecules or particles of contaminant. Organic pollutants, chlorinated aliphatics, and chlorinated pesticides have been effectively removed from wastewater using this method.

✔ **Key Point:** The adsorption method is normally reversible. It is therefore common to remove the adsorbed contaminants after the adsorption capacity of the carbon has been exhausted. This regenerates the carbon, allowing it to be reused.

Aeration

For relatively volatile chemicals, aeration can be used to drive (strip) the contaminants out of solution. These stripping methods typically use air, though in some applications steam is used. The volatiles removed in an air stripper are usually discharged directly into the atmosphere. When discharge into the atmosphere is unacceptable, a GAC treatment method can be added to the exhaust air.

CHEMICAL TREATMENT METHODS

Chemical treatment methods include neutralization, precipitation, solidification-stabilization and reduction-oxidation. Chemically treating hazardous waste has the potential advantage of converting it to less hazardous forms; it can, in some circumstances, produce useful by-products.

Neutralization

In the neutralization method, an acid or base is added to a hazardous waste to achieve pH of 7, resulting in less hazardous and more suitable waste for additional treatment.

Precipitation

A common method of removing heavy metals from a liquid waste is via *chemical precipitation*—a method in which all the substances of a solution are transformed into a solid phase. By properly adjusting pH, settling and filtration can decrease the solubility of toxic metals.

Solidification-Stabilization

Solidification of hazardous waste is commonly practiced before waste is disposed of in landfills. It is a process in which various materials are added to a liquid or semiliquid waste to produce a solid, monolithic mass as end product. *Physical* stabilization is a process in which waste in sludge or semisolid form is mixed with a bulking agent such as pulverized ash, to produce solids, so that it becomes easy to transport in a final disposal method. *Chemical* stabilization is a process in which hazardous wastes react chemically to form insoluble compounds in a stable crystal lattice.

Reduction-Oxidation (Redox)

Reduction-oxidation (redox) reactions are those reactions in which the oxidation state of component atoms changes as a result of the transfer of electrons from one chemical species to another. When electrons are removed from an ion, atom, or molecule, the substance is oxidized; when electrons are added, it is reduced.

✔ **Key Terms:** Wastes that can be treated via redox include benzene, phenols, most organics, cyanide, arsenic, iron, and manganese; those that can be successfully treated using reduction treatment include chromium (VI), mercury, lead, silver, chlorinated organics like PCBs, and unsaturated hydrocarbons.[5]

Additional Reading

Gano, L. *Hazardous Waste.* (Gale Group, 1991).
LaGrega, M. D., et al. *Hazardous Waste Management,* 2nd ed. (New York: McGraw Hill Company, 2000).
Lipeles, M., II. *Hazardous Waste,* 3rd ed. (Anderson Publishing Company, 1997).

Summary

- Despite the fact that chemistry is absolutely essential to our standard of living, it has a poor image with many people. It is often blamed for many problems perceived to adversely affect human health and the environment. These public concerns include many of the environmental problems we have discussed to this point. Many environmental problems—acid rain, global warming, cancer-causing pollutants, air pollution, water pollution, soil pollution—are perceived to be related to the improper handling, storage, and disposal of hazardous waste. Because of such perceptions, human activities involving chemicals have become highly regulated by federal, state, and local laws.

- A hazardous waste is any waste or combination of wastes that poses a substantial present or potential hazard to human health or living organisms because such wastes are nondegradable or persistent in nature or because they can be biologically magnified, or because they can be lethal, or because they may otherwise cause or tend to cause detrimental cumulative effects.
- Under RCRA, hazardous wastes are required to have "cradle-to-grave" tracking of their generation, transportation, treatment, storage, and disposal.
- RCRA lists two main types of hazardous waste: "characteristic" wastes and "listed" wastes.
- "Characteristic" hazardous waste exhibits one of four hazardous characteristics—ignitable, corrosive, reactive (or explosive), or toxic—that are not listed or otherwise excluded in Subtitle C regulation.
- An ignitable hazardous waste bears USEPA number D001.
- A corrosive waste has USEPA number D002.
- A reactive waste has USEPA number D003.
- Toxic characteristic (TC) hazardous wastes are those solid wastes given USEPA numbers D004 through D043.
- The inclusive listing adopted by USEPA includes separate lists designated F, K, P, and U.
- The F List consists of 28 different waste streams from nonspecific sources.
- The K List includes 116 hazardous wastes consisting mostly of residues from specific types of chemical reactions or various distillation or purification processes.
- The P List includes 239 chemical substances that have been identified as acutely hazardous, although they also may be reactive or exhibit other characteristics.
- The U List includes 521 substances identified as toxic wastes.
- USEPA also lists "Exceptions to Hazardous Waste Designation."
- Treatment methods are often categorized as being chemical, physical, biological, thermal, or stabilization/fixation.
- Physical treatment methods are used to separate phases and components in a liquid waste stream.
- Chemical treatment methods include neutralization, precipitation, solidification-stabilization, and reduction-oxidation.

New Word Review

Characteristic hazardous wastes—wastes that exhibit one of four hazardous characteristics: ignitable, corrosive, reactive (or corrosive), or toxic.

Chemical treatment methods—neutralization, precipitation, solidification-stabilization, and reduction-oxidation.

Corrosivity—a characteristic used to classify waste with pH equal to or less than 2.0 or equal to or greater than 12.5.

Hazardous waste—a waste that is (1) listed in Title 40, Part 261 of RCRA, or (2) exhibits any one of the four characteristics—ignitability, corrosivity, reactivity, or toxicity.

Hazardous waste treatment methods—physical, chemical, biological, thermal, or stabilization/fixation.

Ignitability—a substance's susceptibility to burning or catching on fire.

Listed hazardous waste—chemical substances listed under Title 40, Part 261 of the RCRA regulations. In RCRA, hazardous wastes have been placed on one of three lists: nonspecific source wastes, specific source wastes, and commercial chemical products.

Physical treatment methods—ways of separating phases and components in a liquid waste stream.

Reactivity—a characteristic used to classify a hazardous waste that is extremely unstable.

Reduction-oxidation (redox)—those reactions in which the oxidation state of component atoms changes as a result of the transfer of electrons from one chemical species to another.

Toxicity—the degree of danger posed by a toxic substance; a characteristic used to classify a hazardous waste.

Chapter Review Questions

14.1.   When a hazardous material is spilled or accidentally released to the environment, it becomes a _____.

14.2.   Under RCRA, hazardous wastes are required to have _____ tracking of their generation, treatment, storage, and disposal.

14.3.   What characteristic of hazardous waste is designated by the code D003? _____

14.4.   SW846 Test Method 1110 is used to test for _____.

14.5.   The TCLP test is used to test for _____.

14.6.   What contaminant does EPA HW No D014 indicate? _____

14.7.   What consists of 28 different waste streams from nonspecific source? _____

14.8.   _____ include gravity separation, phase change systems, and various filtering operations.

14.9.   _____ is a phase separation method where suspended particles settle under the influence of gravitational forces.

14.10. When electrons are removed from an ion, the substance is _____.

Notes

1. F. R. Spellman, *The Science of Environmental Pollution* (Lancaster, Pa.: Technomic Publishing Company, 1999), 244.

2. USEPA. *The Code of Federal Regulations,* Title 40, Protection of Environment (40 CFR), Revised as of 1995.

3. The complete test for Test Method can be found on the USEPA website at http://epa.gov/epaoswer/hazwaste/test/sw846htm.

4. G. M. Masters, *Introduction to Environmental Engineering and Science* (Englewood Cliffs, N.J.: Prentice-Hall, 1991), 183.

5. USEPA, *Technology Screening Guide for Treatment of CERCLA Soils and Sludges* (Washington, D.C.: USEPA, Office of Solid Waste and Emergency Response, EPA/502/2–88/004, 1988).

CHAPTER 15

Sampling and Analysis

Sampling and analysis constitute a cornerstone of the environmental profession, which encompasses regulatory compliance and policy, science and technology, training, auditing, emergency response, and remedial action.

Along with fundamental (in many cases advanced) training in chemistry, environmental professionals must be well grounded in sampling techniques and analysis. In one situation, the sampling might be rather simple and straightforward. For example, the environmental professional may be called on to sample an underground vault (confined space) to ensure that it is safe to enter. The sampler in this situation typically tests the internal atmosphere to ensure that there is enough oxygen, that the flammability limit is not exceeded, and that toxics are not present. This type of sampling gives immediate results. In a different situation, the sampler may be called upon to sample for chemical contamination in a stream, lake, or river, for example, where results may not be immediate; laboratory analysis may be required.

As mentioned, along with being well versed in proper sampling techniques, the sampler may also be required to analyze the samples gathered. However, in many cases, trained laboratory chemists or technicians who have extensive training in sample analysis perform the actual analysis. Even when this is the case, the sampler may be called upon to interpret the analyst's results.

It is axiomatic that the analytical results obtained in the laboratory can never be more reliable than the sample upon which the tests are performed. Experience has shown that more results are in error because of inadequate sampling than because of faulty laboratory techniques.

In this chapter, I describe the elements of field sampling, laboratory analysis, and interpretation of results. I also briefly describe advanced sampling and analysis using chromatography and spectroscopy and good lab procedures.

Sampling

The key elements involved in sampling include:

- Sampling plans
- USEPA SOPs
- Representative samples
- Discrete and composite samples
- Background samples
- Blanks, splits
- Chain of custody
- Shipping requirements

SAMPLING PLANS

Whenever offsite sampling is required, a *sampling plan* is usually developed by a consultant/contractor. Whoever is hired to develop the sampling plan also defines the scope of all field activities. The analytical chemist or designated lab person (the client), whose daily work site is the laboratory, along with the applicable regulatory agency, reviews/approves the sampling plan.

✔ **Key Point:** Environmental professionals are often qualified to develop their own sampling plans.

When planning a study that involves environmental sampling, it is important, before initiating the study, to determine the objectives of environmental sampling. One important consideration is to determine whether sampling will be accomplished at a single point or at isolated points. Additionally, frequency of sampling must be determined. That is, will sampling be accomplished at hourly, daily, weekly, monthly, or even longer intervals? Whatever sampling frequency is chosen, the entire process will probably continue over a protracted period (i.e., preparing for environmental sampling in the field might take several months from the initial planning stages to the time when actual sampling occurs). One thing is certain; the environmental professional should be centrally involved in all aspects of planning.

USEPA points out that the following issues should be considered in planning the sampling program:[1]

- Availability of reference conditions for sampling area
- Appropriate dates to sample in each season
- Appropriate sampling gear
- Sampling station location
- Availability of laboratory facilities
- Sample storage
- Obtaining a USGS topographical map
- Becoming familiar with health and safety procedures

Once the initial objectives (issues) have been determined, the plan is devised. The written plan should spell out:

- Health and safety procedures
- Field sampling SOPs
- Number/type/location/depth of samples
- Test methods
- Chain of custody
- QA/QC procedures

To expedite rapid reaction to changing field conditions, sampling plans should *always* specify the chain of command and multiple points of contact for both the consultant and the client.

✔ **Key Point:** Consider attaching to all sampling plans a contact sheet that summarizes contact names and expected reporting frequency for all parties involved.

✔ **Key Point:** Once the initial objectives (issues) and procedures for the sampling are written, the sampler can move on to other important aspects of the sampling procedure. Obviously, along with the items just listed, it is imperative that the sampler understand what environmental sampling is all about.

USEPA SOPS

Some USEPA Regions have developed standard operating procedures (SOPs), which they expect to be followed whenever USEPA oversees field activities. These SOPs typically spell out accepted field sampling techniques, with special emphasis on procedures such as:

- Decontamination of sampling equipment
- Disposal of purge waters
- Well installation and development
- Sample filtration (where allowed)

REPRESENTATIVE SAMPLES

A representative sample is a sample containing all the constituents present in the media or waste being sampled. USEPA will not accept analytical results from samples that are not representative. Homogeneous materials are relatively straightforward to sample, but nonhomogeneous substances may require multiple samples to fully characterize.

✔ **Key Point:** Regulatory agencies generally have the right to collect their own "representative sample" to determine if they concur with a generator's waste identification.

DISCRETE AND COMPOSITE SAMPLES

When the nature, distribution, and extent of contamination is not well known, the collection of "discrete" (i.e., individual) samples is generally recommended.

In cases where the contamination is better characterized and/or uniform, it may be appropriate (and much cheaper) to mix samples from different locations into one composite. The term *composite sample* refers to a combination of discrete (grab) samples collected at the same sampling point at different times. When collecting composite samples, the generator should be able to demonstrate that the mixing of multiple samples will not dilute or otherwise skew the analytical results. USEPA has provided written guidance on how to collect composite samples.

BACKGROUND SAMPLES

Background samples are collected to establish the natural concentrations of constituents in the local environment prior to the introduction of potential contamination. In many heavily industrialized areas, it is often very difficult to determine true "background" conditions characterizing any of the natural concentrations in any of the three environmental mediums (i.e., air, water, and soil). To avoid these problems, a "background" sample is sometimes collected from an area, water body, or site "similar" to one being studied but located in an area known or thought to be free from pollutants of concern.

BLANKS, SPLITS

To assure the quality of sampling results, good sampling plans generally call for the collection of *trip blanks* and a variety of *field blanks*. USEPA Standard Procedures require quality control (QC) samples be collected in addition to the objective samples taken during field activities, including investigations, and studies. The number and type of QC or quality analysis (QA) samples are specified to provide evidence that the objective samples

- are representative of the materials sample, and that
- samples were handled, transported, and analyzed in a manner that provides accurate, precise results.

✔ **Key Point:** QA samples that serve to provide evidence that the objective samples are representative of the materials being sampled include background samples for comparison, splits (see below) to define handling variability, and duplicate samples of the materials being sampled as defined by EPA SOPs.

A *trip blank* is a sample that is prepared prior to the sampling event in the actual container and is stored with the investigative samples throughout the sampling event. It is packaged for shipment with the other samples and submitted for analysis. At no time after their preparation are trip blanks to be opened before they reach the laboratory. Trip blanks are used to determine if samples were contaminated during storage and/or transportation back to the laboratory (a measure of sample handling variability resulting in positive bias in contaminant concentration). If samples are to be shipped, trip blanks are to be provided with each shipment but not for each cooler. Trip blanks are required only with sampling for volatile organic contaminants.

An *equipment field blank* (rinse blank) is a sample from organic-free water that has been run over/through sample collection equipment. These samples are used to determine if contaminants have been introduced by contact of the sample medium with sampling equipment. Equipment field blanks are often associated with collecting rinse blanks of equipment that has been field cleaned. Rinse blanks should be collected daily, or in some instances can be collected for a decontaminated set of equipment (multiple soil samples) that were all prepared (decontaminated) at one time.

A *field blank* is a sample that is prepared in the field to evaluate the potential for contamination of a sample by site contaminants from a source not associated with the sample collected (for example, airborne dust or organic vapors that could contaminate a soil sample). Organic-free water is taken to the field in sealed containers or generated on-site. The water is poured into the appropriate sample containers at predesignated locations at the site. Field blanks should be collected in dusty environments and/or from areas where volatile organic contamination is present in the atmosphere and originating from a source other than the source being sampled. One or two field blanks are generally adequate for a field exercise lasting a few days. Additional field blanks should be collected if field conditions change (i.e., high winds and dust creating unrelated field contamination).

To double-check the accuracy of analytical results (above and beyond normal laboratory QA/QC procedures), a single sample is often split. A *split sample* is a sample that has been portioned into two or more containers from a single sample container or sample-mixing container. The primary purpose of a split sample is to measure sample handling variability. Split samples may also be used to evaluate and compare the performance of two (or more) analytical methods, or laboratories. Widely disparate results from a single split sample point to errors in laboratory procedures rather than problems in the field.

▶ **Key Point:** Planning QC sampling requirements should be included in a sampling plan objectives workup that is developed prior to the field sampling activities.

CHAIN OF CUSTODY (COC)

A sampling plan requires very detailed chain-of-custody (COC) procedures to ensure sample integrity from collections to data reporting—that is, to assure that samples are not tampered with. A COC form must follow all sample shipments to record the times, locations, and names of all persons taking possession of samples.

The following procedures summarize the major aspects of a COC form to accompany each sample or group of samples. More detailed discussions are available.[2]

Sample labels: Special labeling is used to prevent sample misidentification. The following information should be included on the label: sample number, name of collector, date and time of collection, place of collection, and sample preservative. Use waterproof ink to provide the label information.

Sample seals: Sample seals are used to detect unauthorized tampering with samples up to the time of analysis. Seals should be attached in such a way that it is necessary to break it to open the sample container.

Field logbook: All information pertinent to a field survey or sampling should be recorded in the field logbook. Recorded information should include purpose of sampling, location of sampling point, name and address of field contact, producer of material being sampled, and address.

Sample analysis request sheet: The sample analysis request sheet should accompany the sample to the laboratory.

Sample delivery to the laboratory: Samples should be delivered to laboratory as soon as practicable after collection, typically within two days.

Receipt and logging of sample: In the laboratory, the sample custodian inspects the condition and seal of the sample, reconciles label information and seal against the COC record, assigns a laboratory number, logs the sample in the laboratory log book, and stores it in a secured storage room or cabinet until it is assigned to an analyst.

Assignment of sample for analysis: The laboratory supervisor usually assigns the sample for analysis. Once the sample is in the lab, the supervisor or analyst is responsible for its care and custody.

SHIPPING REQUIREMENTS

RCRA provides a hazardous waste exclusion for all samples being shipped to and from laboratories *if* they meet all requirements. Many waste and product samples *cannot* be legally shipped by air, and there are many stringent labeling and packaging requirements that must be followed.

Laboratory Analysis

Laboratory analysis typically consists of two important subparts: qualitative and quantitative analysis. *Qualitative analysis* is the determination of the identities, but not the concentrations, of the constituents present in a substance, a mixture of substances, or a solution. The process of determining the concentrations of the species present or the percent composition of a mixture is called *quantitative analysis.*

In this section I discuss the key components involved in both qualitative and quantitative analysis. Specifically, we describe detection limits, common analytical procedures, and QA/QC (i.e., spike recoveries, data validation, and comparison to Standards).

DETECTION LIMITS

Detection limits (DLs) refer to a minimum concentration of an analyte that can be measured above the instrument background noise. Thus, when detection limits are used as reporting limits, the lab is saying that the analyte is not present at or above the value given. It may be present at a lower concentration, but it cannot be "seen" by the instrument.

Method detection limits (MDLs) refer to the lowest concentration that can be detected by an instrument with correction for the effects of sample matrix and method-specific parameters such as sample preparation. MDLs are explicitly determined as set forth in 40 CFR Part 136. They are defined as three times the standard deviation of replicate spiked analyses.

Practical quantification limit (PQL) refers to the practical concentration that can be reliably achieved within specified limits of precision and accuracy during routine laboratory operating conditions. PQLs normally are arbitrarily set rather than explicitly determined. Most organic SW846 methods give PQLs (prior to 1994). The SW846 PQLs are arbitrarily set at some multiple of typical MDLs for reagent water. Multiplying factors are given for various matrices such as groundwater, wastewater, soil, and sludge, etc. Generally, laboratories use the SW846 PQLs (adjusted for sample size, dilution, and % moisture) for reporting limits, but they may use PQLs that they have generated. SW846 does not stipulate how to handle organic analytes that are positively identified at a concentration below the SW846 PQL. Generally, labs *do not* report these as present.

✔ **Key Point:** Ensure that the lab provides a numeric value for the detection limit. Do not accept "BDL" (below detection limit)—insist on a real number. Check to be sure that this limit is acceptable; for example, is the specified DL below the regulatory level for the contaminant(s) of concern?

COMMON ANALYTICAL PROCEDURES

Environmental professionals involved in sampling and analysis are primarily concerned with analytical procedures focused on those substances that contaminate the workplace and/or environment. They would normally be acquainted with a variety of such procedures, of which the following are the most common.

VOCs and Semi-Vols (Volatiles). There are specific test methods, depending on which regulations apply (e.g., Volatile Organics, USEPA 8240, 8260). Holding times and handling procedures are very important.

TOC (total organic carbon). TOC is amount of covalently bonded C in a water sample; the sample is burned and the resulting CO_2 is measured.

TPH (total petroleum hydrocarbons). A procedure typically required in underground storage tank (UST) investigations. Holding times and handling procedures are very important. Modified EPA 8100 analyzes for hydrocarbons as kerosene, diesel fuel, heavy oils, mineral spirits, varsol, and fuel oil (DL = 10 mg/kg). Modified EPA 8015 analyzes for hydrocarbons as gasoline (DL = 10 mg/kg).

BTEX (benzene, toluene, ethylbenzene, xylenes). A test for aromatic volatile organics (EPA 8020), DL = 500 µg/kg (see figure 15.1). Holding times and handling procedures are very important.

PAHs (polyaromatic hydrocarbons). These compounds are formed by the incomplete combustion of hydrocarbons. Sources include engine exhausts, wood stove

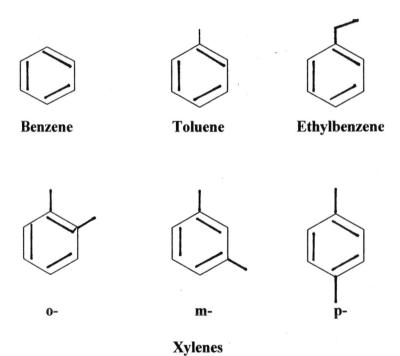

Benzene   **Toluene**   **Ethylbenzene**

o-   **m-**   **p-**

Xylenes

Figure 15.1.   BTEX (benzene, toluene, ethylbenzene, and xylenes).

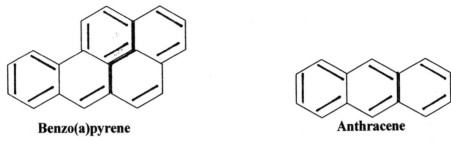

Benzo(a)pyrene   **Anthracene**

Figure 15.2. PAHs.

smoke, cigarette smoke, asphalt, etc. They are the most significant atmospheric organic particles and are also abundant in soils. The high temperatures within internal
combustion engines (>500°C) break C-H and C-C bonds to form free radicals. These
free radicals recombine to form complex, thermally resistant aromatic ring structures
(see figure 15.2).

Pesticides/herbicides (see figure 15.3). Halogenated, N- and P-based pesticides and
herbicides are analyzed in EPA 500, 600, and 8000 series.

PCBs (arochlor)(see figure 15.4). PCBs (polychlorinated biphenyls) are also analyzed by EPA test methods series 500, 600, and 8000.

Total metals. Various test methods and analytical techniques are used, depending
on type of sample (soil, water, etc.), required DL, and applicable regulations. Total
metals analysis includes the 8 RCRA metals: As, Cd, Pb, Hg, Ag, Ba, Se, Cr. For
testing lead in water, use EPA Method 6010.

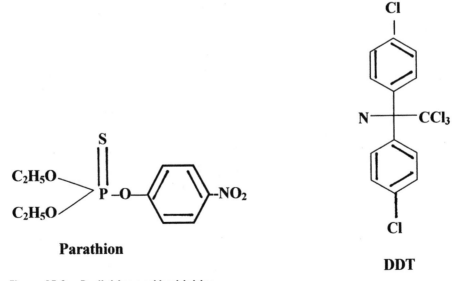

Parathion

DDT

Figure 15.3.   Pesticides and herbicides.

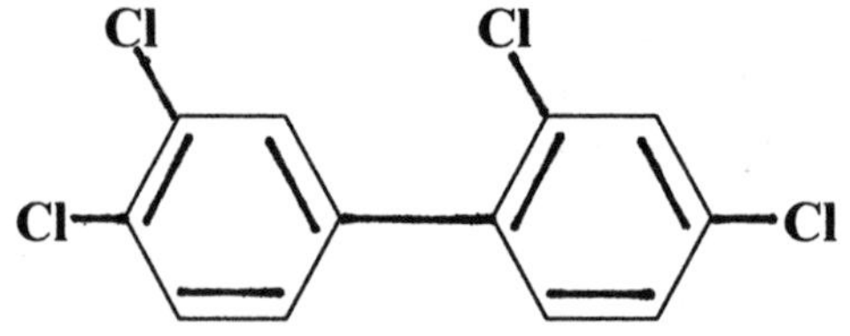

4-Cl PCB isomer

Figure 15.4. PCBs (arochlor).

Atomic absorption (AA). This procedure can also be used to determine total metals in water samples. Each metal has a characteristic *absorption* pattern when heated; the degree of absorption is proportional to the amount of metal present in the sample.

ICP/MS (Inductively Coupled Plasma/Mass Spectrometry) can measure metal content in a wider variety of samples (water, soil, air, etc.) with less interference. Each metal has a characteristic *emission* pattern when heated to very high temperatures. The intensity of the emission is proportional to the amount of metal present.

▶ **Key Point:** If only one metal is being analyzed, AA can be more cost effective. ICP/MS (mass spectrometry) can simultaneously determine up to 80 elements in a single can with very low detection limits (down to ppt).

TCLP (Toxicity Characteristic Leaching Procedure). This analytical procedure is used to determine if a waste should be assigned any Toxicity Characteristic waste codes (D004–D043). The procedure produces a "model leachate" that is supposed to assess the leachability of toxic constituents in the waste. A full TCLP for all 40 constituents can be requested, or a partial analysis can be completed for one or more individual constituents (e.g., lead).

Flash point (ignitability). The standard "closed cup" test method is available to determine the temperature at which a liquid flashes.

▶ **Key Point:** The definition of "ignitability" depends on the applicable regulations.

pH (corrosivity). This is a simple test with a calibrated pH meter for aqueous materials (see figure 15.5).

BOD/COD (Biological Oxygen Demand/Chemical Oxygen Demand). BOD measures the amount of oxygen that would be consumed if all the organics in a water sample were oxidized by microorganisms. The test takes 5 days to complete (*BOD5*).

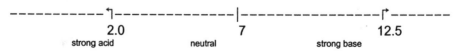

Figure 15.5. The pH scale.

An average clear unpolluted lake sample might measure 2 mg/L. A raw sewage sample might measure >1,000 mg/L. COD measures the amount of oxygen consumed in the chemical oxidation of all oxidizable organic and inorganic matter in a sample. This test can be completed in two hours.

TSS/TDS (Total Suspended/Dissolved Solids). TSS/TDS are those particles which can be removed from a solution by filtration (usually ≥ 0.45 micron). TSS are measured by evaporating the filtered sample to determine the dried weight of the dissolved particles in mg/L. Conductivity measurements are usually proportional to TDS measurements, since most of the dissolved solids are usually salts (ions).

Turbidity. This test measures the amount of solid matter suspended in water, based on the amount of light it scatters. It is measured in nephelometric turbidity units (NTU). Potable water should not exceed 0.5 NTU.

Quality Assurance and Quality Control (QA/QC)

Laboratories generally follow standard Quality Assurance and Quality Control (QA/QC) procedures. Before contracting services from a laboratory, their Quality Assurance Plan should be reviewed or proof of accreditation should be requested.

Quality assurance (QA) activities provide a formalized system for evaluating the technical adequacy of sample collection and laboratory analysis activities. QA activities begin before samples are collected and continue after laboratory analyses are completed, requiring ongoing coordination and oversight. The QA program should integrate management and technical practices into a single system to provide data that are sufficient, appropriate, and of known and documented quality.

Quality control (QC) practices consist of more focused, routine, day-to-day activities carried out within the scope of the overall QA program. QC is the routine application of procedures for obtaining data that are accurate (precise and unbiased), representative, comparable, and complete. QC procedures include activities such as identification of sampling and analytical methods, calibration and standardization, and sample custody and record keeping. Audits, reviews, and complete and thorough documentation are used to verify compliance with predefined QC procedures.

SPIKE RECOVERIES

Spike recoveries are analyzed to provide a measure of the accuracy of the methods used for analyses. Standard QA/QC analytical procedures include the addition (i.e., the spiking) of a known quantity of the contaminant of concern to determine the percent recovery. If the percent of recovery is low, then there may be a high contaminant concentration and yet a "low" analytical result, since the instrument or analytical procedure is not accurately detecting this constituent.

✔ **Key Point:** Spike recoveries are routinely determined as part of any good laboratory QA/QC. Always check to see that the percent recoveries are not so low as to invalidate the data.

DATA VALIDATION AND BASIC STATISTICS

An important part of QA/QC is data validation, a detailed analysis of a data package to determine data quality by checking such parameters as:

• Holding times
• Calculations performed by the laboratory
• Spike recoveries and associated calculations
• Unit analysis, data transportation, etc.

When data quality is extremely important, consider requesting a *data validation package* from the laboratory, especially if

• the data will be scrutinized in court;
• there have been random but persistent problems with data quality; and/or
• regulatory agencies request such data.

✔ **Key Point:** Laboratory data should be compared to standards. Request that the laboratory provide all applicable regulatory levels for the constituents being analyzed. If standards are not applicable, compare the data to background levels, or establish an alternate cleanup level with the appropriate regulatory agencies.

BASIC STATISTICAL CONCEPTS

In order to interpret the data analysis information, it is important to check the accuracy of the results. Probably the most important step in this process is the *statistical analysis* of the results. Thus, a very basic understanding of statistical methods is required.

The principal concept of statistics is that of variation. In conducting a sampling protocol for environmental contaminants, variation is commonly found. Variation comes from the methods that were employed in the sampling process or in the distribution of organisms. Several complex statistical tests can be used to determine the accuracy of data results. In this basic discussion, however, only basic calculations are presented.

The basic statistical terms include the *mean* or *average*, the *median*, the *mode*, and the *range*. The following is an explanation of each of these terms.

Mean—the total of the values of a set of observations divided by the number of observations.

Median—the value of the central item when the data are arrayed in size.

Mode—the observation that occurs with the greatest frequency and thus is the most "fashionable" value.

Range—the difference between the values of the highest and lowest terms.

Given the following laboratory results for the measurement of dissolved oxygen (DO), find the mean, median, mode, and range. The data are: 6.5 mg/L, 6.4 mg/L, 7.0 mg/L, 6.9 mg/L, 7.0 mg/L. To find the mean:

$$\text{Mean} = \frac{\begin{array}{c}(6.5 \text{ mg/L} + 6.4 \text{ mg/L} + 7.0 \text{ mg/L} \\ + 6.9 \text{ mg/L} + 7.0 \text{ mg/L})\end{array}}{5} = 6.58 \text{ mg/L}$$

To find the mode:

$$\text{Mode} = 7.0 \text{ mg/L (number that appears most often)}$$

To find the median:

$$\text{Median} = 6.9 \text{ mg/L (central value)}$$

To find the range:

$$\text{Range} = 7.0 \text{ mg/L} - 6.4 \text{ mg/L} = 0.6 \text{ mg/L}$$

The importance of using statistically valid methods cannot be overemphasized. Several different methodologies are available. A careful review of the methods available should be made before computing analytic results. Using appropriate sampling methods along with careful lab analysis will provide data that is accurate.

✔ **Key Point:** The need for statistics in environmental sampling is driven by the science itself. Environmental studies often deal with entities that are variable. If there were no variation in environmental data, there would be no need for statistical methods. In this text we have only scratched the surface in revealing the basics. Because analytical data may be provided with standard deviation and variance information, environmental professionals should have further training on the knowledge of these terms, how to compute them, and how to interpret them.

Advanced Sampling and Analysis

Today's modern environmental laboratories are equipped with the latest technological tools. We can perform very detailed and accurate analysis on environmental samples today that could hardly be imagined just a few decades ago. In this section, I briefly describe two of these advanced technologies, chromatography and spectroscopy.

Both chromatography and spectroscopy are relatively modern analytical techniques designed to identify and sometimes also purify organic compounds. Prior to the advent of these modern methods, chemists were forced to rely almost entirely on elemental analysis and chemical reactivity to deduce molecular structure.

CHROMATOGRAPHY

Chromatography is an extremely versatile method of separating solids, liquids, or gases that was developed in the 1950s. The technique was named from the Greek *chroma,* meaning color, because the early uses of this technique identified the components of the separated mixture by color.

In general, chromatographic methods involve passing a liquid or gaseous mixture of compounds across a solid matrix to take advantage of the different "affinities" of each compound for the solid matrix.

✔ **Key Point:** By changing the matrix and/or the solvent, the differences in affinity can be maximized.

There are many different types of chromatography, including:

• Paper chromatography
• Thin layer chromatography (TLC)
• Column and preparative chromatography
• Liquid chromatography (LC)
• High-performance liquid chromatography (HPLC)
• Gas chromatography (GC)

Paper chromatography is a simple way of illustrating the basic chromatographic technique (see figure 15.6).

1. A tiny drop of the compounds to be separated is "spotted" onto a very small piece of special paper (-2 cm $\times$ 7 cm).
2. This paper is then placed in a small covered breaker containing a suitable organic solvent/solvent mixture, so that the long end of the paper is vertical and the "spot" is just *above* the solvent line.

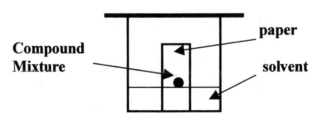

Figure 15.6. Paper chromatography.

3. The paper is left undisturbed until the solvent is drawn up near the top of the paper by capillary action.
4. The paper is then removed from the solvent and sprayed with a suitable developing agent to identity the presence of the separated species.

To improve the separation, the length of the paper can be increased and/or the developing solvent can be changed. The developing agent can be a colorimetric agent and/or a chemically reactive species that forms colored compounds with the types of molecules to be identified.

- 2,4-dinitrophenylhydrazine turns red to yellow in the presence of aldehydes and ketones.
- Ninhydrin turns blue/purple in the presence of amino acids.

Thin layer chromatography (TLC) is exactly the same as paper chromatography, except that the solid matrix is typically a thin coating of silica gel, alumina, cellulose, or magnesium silicate. Commercial thin layer chromatography "plates" are usually produced on a thin, stiff layer of aluminum foil that can be cut to the desired size.

Preparative chromatography is identical to thin layer chromatography but on a much larger scale (see figure 15.7). It is designed to separate and recover larger quantities of material for further experimentation or analysis.

Column chromatography is exactly the same as preparative chromatography, except that the separation is performed in a glass column that is *eluted* with a suitable solvent (see figure 15.8).

Liquid chromatography (LC) follows exactly the same principle as column chromatography, except on a much smaller scale in which the "column" is a tube filled with a solid matrix.

High-performance liquid chromatography (HPLC) is used to efficiently separate extremely small quantities of compounds into extremely pure forms for further analysis.

Gas chromatography (GC) is used to separate volatile organics in the *vapor* state. Otherwise it is similar to LC and HPLC in that the solid matrix is packed into tubing.

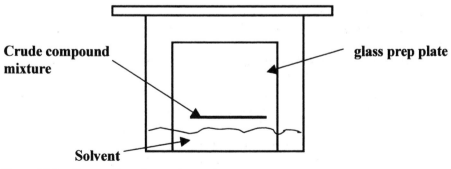

Figure 15.7. Preparative chromatography.

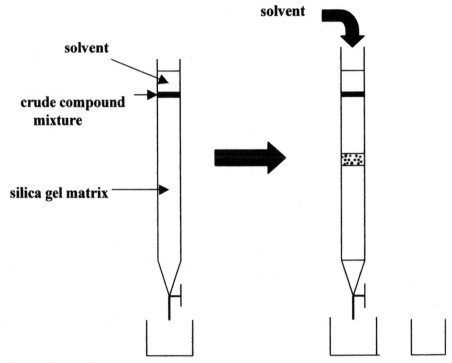

Figure 15.8.   Column chromatography.

GC analysis can be coupled to a *mass spectrometer* (GCMS) to identify the separated compounds by the molecular weight of their fragments.

SPECTROSCOPY

Spectroscopy[3] encompasses a broad range of techniques used to identify compounds generally based on the way in which they interact with *radiant energy*. The energy of the radiation absorbed by a compound depends on its structure. The type of radiant energy can be described by its wavelength.

The *electromagnetic spectrum* describes the full range of *wavelengths* and *energies* of radiant energy (see figure 15.9).

Different types of spectroscopy use different types of radiant energy to elucidate the structure of compounds, as shown in table 15.1.

Infrared spectroscopy (IR) helps chemists identify which functional groups are present (e.g., -OH, C=O, C-O). It relies on the fact that compounds can absorb only those wavelengths of infrared radiation which have exactly the right energy to increase the vibrations of one or more chemical bonds (see figure 15.10).

When certain wavelengths of infrared radiation are absorbed, the *intensity* of the radiation at that wavelength decreases (see figure 15.11). An IR *spectrometer* (or *spectro-*

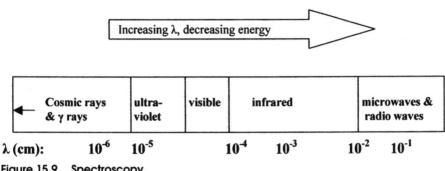

Figure 15.9.   Spectroscopy.

Table 15.1.   Spectroscopy and radiant energy.

Type of Spectroscopy	Type of Radiation	Molecular Changes Induced
Infrared (IR)	$\lambda = 2.5$–$16~\mu m$	Bond excitation
Ultraviolet (UV)	$\lambda = 200$–$750~nm$	e- to higher energy orbital

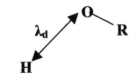

Figure 15.10.   Infrared spectroscopy.

Figure 15.11.   The intensity of radiation decreases.

photometer) measures this change in intensity per wavelength as *percent transmission* (figure 15.12).

Ultraviolet (UV) spectroscopy helps chemists to identify the quantity of a compound and the presence of double and triple bonds. It relies on the fact that UV light of exactly the right wavelength is sufficiently energetic to promote one or more electrons to a higher energy orbital (see figure 15.13).

As with IR, when certain wavelengths of ultraviolet radiation are absorbed, the *intensity* of the radiation at that wavelength decreases. UV *spectrophotometers* measure this change in intensity per wavelength as *absorbance* (a logarithmic function of the initial intensity divided by the final intensity; see figure 15.14).

UV spectroscopy can determine the *concentration* of a sample by comparing the

Figure 15.12.   Percent transmission.

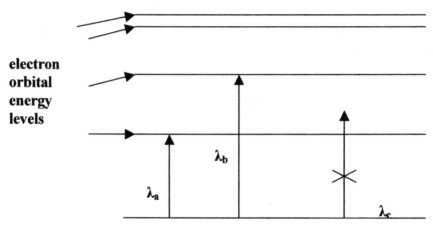

Figure 15.13.   Electron excitation by UV light of the correct wavelength.

absorbance of the sample to a standard curve. Measuring the absorbances of known concentrations of the sample (see figure 15.15) creates these standard "curves." Metals have a characteristic absorption pattern when mildly heated (see figure 15.16).

This characteristic absorption pattern is analogous to a stationary ball sitting on a staircase. This ball can only have specific fixed levels of potential energy (see figure 15.17).

Atomic absorption can be used to determine the total concentration of metals in water samples. The intensity of the absorption is proportional to the amount of metal present in the sample. Metals also have characteristic *emission* patterns when heated to very high temperatures (see figure 15.18).

ICP/MS (inductively coupled plasma/mass spectrometry) can measure metal content in a wider variety of samples (water, soil, air, etc.) with less interference than atomic absorption. The intensity of the emission is proportional to the amount of metal present.

Nuclear magnetic resonance spectroscopy (NMR) helps chemists to identify the *carbon backbone* of a molecule and the presence of *hydrocarbon functional groups*.

NMR spectroscopy relies on the way in which certain nuclei interact with very long wavelength radiant energy (i.e., radio waves). In 1946, the original magnetic resonance phenomenon was first discovered independently by Bloch and Purcell, who received a Nobel Prize in 1952 for their work.

The atomic nuclei of all the isotopes of all the elements can be classified as either

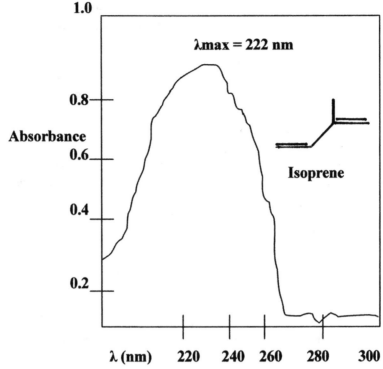

Figure 15.14. Absorbance.

having spin or *not having spin*. NMR helps to define the C backbone and hydrocarbon functionality because ^1H and ^{13}C have spin and ^{12}C and ^{16}O do not have spin.

🖊 **Key Point:** ^1H and ^{13}C can be distinguished from each other because they do not absorb energy at the same wavelength.

A nucleus with a spin gives rise to a small magnetic field whose magnitude and direction is described by a *magnetic moment.* These magnetic moments are randomly oriented in the absence of an external magnetic field (see figure 15.19).

In the presence of a strong external magnetic field, these magnetic moments align parallel and anti-parallel (see figure 15.20). When nuclei in an applied magnetic field are irradiated with radiant energy of the correct wavelength, a few more nuclei absorb energy and "flip" to the higher energy state. At the wavelength where "flipping" occurs, the nuclei are said to be *in resonance.*

Magnetic resonance imaging (MRI) is based on the principles of NMR, but the name was changed in the 1970s to prevent the general public from becoming alarmed by the negative connotations of the word *nuclear.* Because of the high content of fat

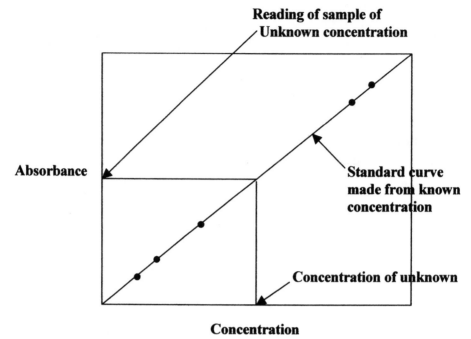

Figure 15.15.   Measuring the absorbance of a sample to a standard curve.

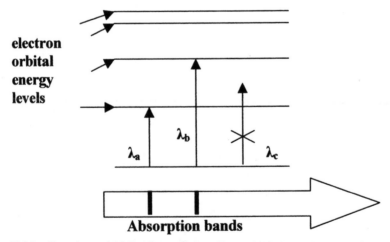

Figure 15.16.   The characteristic absorption pattern of mildly heated metals.

Figure 15.17. Absorption spectra.

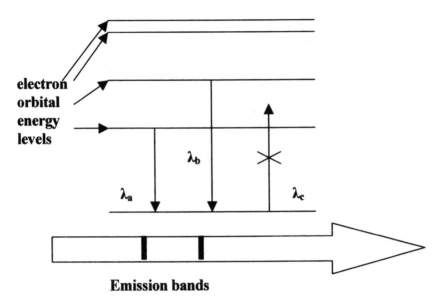

Emission bands

Figure 15.18. Emission spectra.

and water, the human body consists of approximately 63% hydrogen atoms. By perfecting various computerized imaging techniques, the multitude of ^{1}H resonances in the body can be transformed into images of the tissues in which they are found.

Mass spectroscopy (MS, but often known as "mass spec") is based on different principles from the techniques previously discussed. A mass sample is vaporized and then bombarded with electrons of sufficient energy to ionize the compound of interest. These organic ions then fall apart into predictable fragments that are characteristic of the original compound's structure.

A *mass spectrometer* detects the positively charged ions/fragments, recording both their mass and their relative abundance. The pattern of fragments is complex, but the overall spectrum serves as a "fingerprint" for that particular molecule. Mass spectrum of methanol (CH_3OH) is shown in figure 15.21.

The simplest ionization is the loss of one electron, forming the *molecular ion peak* as shown in figure 15.22. The *largest* peak in the spectrum is named the *base peak* and

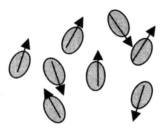

Figure 15.19.   Randomly aligned nuclear magnetic moments.

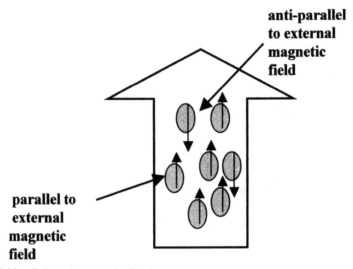

anti-parallel to external magnetic field

parallel to external magnetic field

Figure 15.20.   External magnetic field.

is always assigned an intensity of 100% (see figure 15.23). The remaining significant peaks are derived as shown in figure 15.24.

Working with mass spectroscopy data is like putting together the pieces of a puzzle knowing the overall size of the puzzle (the molecular ion peak) and some of the characteristic pieces.

✔ **Key Point:** When MS is coupled with other techniques, such as GC or HPLC, the "combined" clues are usually sufficient to identify the compound without any doubt.

Good Laboratory Practices

Under OSHA's 29 CFR 1910.1450, Laboratory Standard, all laboratory personnel are required to practice prudent chemical handling. The sudden, violent, and unforeseen

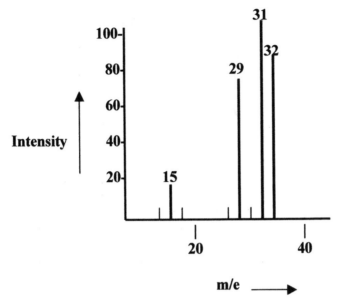

Figure 15.21.   The mass spectrum of methanol (CH₃OH).

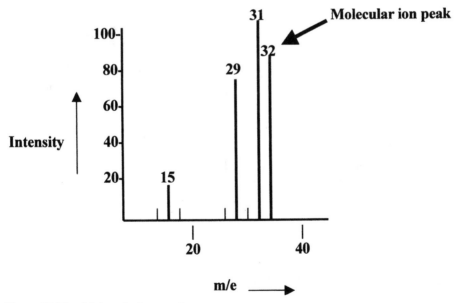

Figure 15.22.   Molecular ion peak.

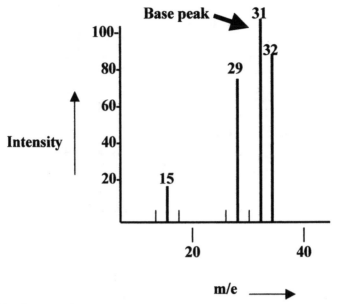

Figure 15.23.   Base peak.

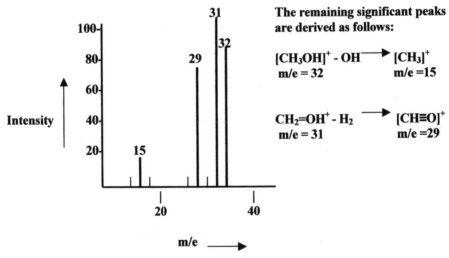

Figure 15.24.   Other significant peaks.

hazards that may result from unsafe chemical handling operations must be guarded against. In order to prevent such mishaps, several safety guidelines should be in place in the laboratory. The purpose of this section is to point out several good laboratory practices, safety guidelines, and in-house chemical spill procedures that should be a significant part of the organizational culture of the environmental laboratory. The following are good practices in the environmental laboratory:

- Use care and caution when working in the laboratory.
- Store chemicals in a safe place where they are not hazardous to personnel or property.
- Label all containers, indicating the chemical name and date of preparation and/or container opening.
- Check labels on all chemical containers before using to ensure that the proper chemicals are selected for intended use.
- Properly dispose of unlabeled or out-of-date chemicals. *Never* dispose of chemicals in common trash containers. When hazardous material is to be disposed of, contact laboratory supervisor to ensure that proper disposal procedures are followed.
- Read directions for each chemical's use and safety. This information is found on the chemical's Material Safety Data Sheet (MSDS).
- Follow directions carefully. *Never* mix chemicals randomly or indiscriminately.
- Handle chemicals carefully when pouring or measuring to prevent spillage.
- Immediately clean up chemical spill according to the directions on the chemical's MSDS.
- Avoid physical contact with chemicals.
- Wear correct gloves for the chemical being handled. Refer to the chemical's MSDS if unsure about proper handling protection. *Never* touch chemicals with bare hands.
- Ensure that protective gloves are free of cracks, tears, and holes and that gloves fit properly before handling chemicals.
- Do not place fingers into mouth, nose, ears, or eyes while handling chemicals.
- Wash hands with a disinfectant soap after handling chemicals.
- Wash off chemical spills on skin immediately with running water.
- Do not smoke or eat in the laboratory while handling chemicals or performing tests.
- Avoid breathing chemical vapors, dusts, or fumes.
- Provide positive ventilation to laboratory work areas.
- Properly dispose of all broken, chipped, or cracked glassware.
- Do not use laboratory glassware as coffee cups or food containers.
- Use a suction bulb to pipette chemicals or wastewater. *Never* use mouth to suck up a fluid or chemical in a pipette.
- As required by the applicable MSDS, use safety goggles and/or a face shield when transferring and measuring chemicals, or whenever there is a potential for chemicals to be splashed in the eyes.
- Use tongs or thermal gloves when handling hot utensils.
- Use only properly grounded electrical equipment.
- Always *add acid to water, not* water to acid.

- Use the carbon dioxide or dry-chemical type of fire extinguishers to control laboratory fires.
- Ensure that the laboratory is equipped with properly working emergency eyewash/ shower. Laboratory workers should anticipate the need to use emergency eyewash/ showers and be familiar with the quickest route to each eyewash/shower station in close proximity to their workstation.
- Ensure that prominent signs warning workers against hot areas such as vents, hot plates, furnaces, water baths, and digestion apparatus are posted in the laboratory.
- Inspect acid-neutralizing tanks and basins that service lab sinks regularly and recharge with correct neutralizing agent when needed.
- Ensure that emergency phone numbers are posted by all telephones.

GENERAL SAFETY GUIDELINES

As with good work practices, general safety guidelines are designed to control chemical hazards in the laboratory. Control of these chemical hazards, of course, should start with engineering controls, followed by administrative controls, and finally by the use of personal protective equipment (PPE). Safety guidelines should be included whenever safety training is conducted for lab workers. Moreover, anytime a standard operating procedure (SOP) is written for a specific lab operation, it is important to include applicable safety guidelines. The following guidelines should be considered for inclusion into the lab's training program and SOPs.

- Only the minimum quantities of chemicals consistent with normal lab needs and requirements should be maintained.
- Before lab personnel receive chemical/hazardous materials, lab management should ensure that workers have been trained on proper handling of chemicals and storage requirements and that no chemical container is accepted without an adequate identifying label and MSDS. In addition, no container of chemical/hazardous material should be accepted unless it is thoroughly inspected to ensure it is damage and leak free.
- Chemicals in storage should be protected to preclude spills, leaks, fire, earthquake damage, etc.
- To minimize the risks associated with chemical incompatibility, the laboratory should establish a segregation scheme for chemicals in storage to ensure that accidental breakage, leaks, or other mishaps to chemical containers do not result if they should react with incompatible materials.
- Storage practices for flammable liquids (flash point less than 100°F) include several guidelines. In general [according to the National Fire Protection Association (NFPA-45)], proper storage practices include the following:
  - The quantities of flammable chemicals stored in the lab should be kept to a minimum.
  - Flammable liquids must be stored away from heat and ignition sources.

- Small quantities (working amounts) of flammable chemicals may be stored on open shelves.
- Flammables should not be stored in areas exposed to direct sunlight.
- Flammable chemicals must be stored in flammable liquid storage cabinets that have been approved by the NFPA. The following safety practices should be observed: (1) store only compatible materials inside the cabinet; (2) follow manufacturers' recommended quantity limits and do not overload cabinets; (3) follow NFPA guidelines for maximum allowable volumes; and (4) do not store paper, cardboard, or other combustible materials in a flammable liquid storage cabinet.
- Flammable liquids in large quantities should be stored in metal safety cans. The cans should be used only as recommended by the manufacturer.
- Peroxides may be explosive, especially whenever they are improperly stored or handled. If peroxides are used in the lab, the following guidelines should be followed:
  - Store peroxide compounds away from heat and light sources.
  - Do not use metal containers because some metal oxides can promote the formation of peroxides.
  - Always label peroxides with the date they were opened.
- Higher chemical exposure risk is associated with lab operations such as weighing, pouring, siphoning, mixing, blending, stirring, and shaking. When performing these operations, care and caution are advised.
- When handling corrosives (acids and bases), wear appropriate personal protective equipment (PPE), including clothing, apron, chemical-resistant gloves, and splash goggles/face shield. Use proper pouring techniques when pouring acids into water and conduct the procedure in a laboratory fume hood. In addition, all dilutions of corrosives must be performed in a lab hood.
- When handling flammables, keep flammable compounds away from any ignition source, such as open flame. Moreover, whenever transferring the chemical from one container to another or heating it in an open container, use a laboratory fume hood.
- Whenever chemicals are transported outside the lab, the container should be placed in a secondary, non-breakable container. Before moving these containers, ensure that the caps are tight.

IN-LAB CHEMICAL SPILL RESPONSE

Lab workers should be aware that required safety training for lab workers includes *emergency response training*. The OSHA Hazard Communication Standard and the OSHA Standard *Hazardous Waste Operations and Emergency Response* (HAZWOPER) mandate such training.

Emergency training applies to building evacuation procedures during fires and explosions, recognition of system alarms, and appropriate action in the event of spills of hazardous materials in the lab. Lab technicians must receive training to distinguish between the types of spills they can handle on their own and those spills that are classified as "major." Major spills dictate the need for outside assistance.

Lab technicians should be qualified to clean up spills that are "incidental." OSHA defines an incidental spill as a spill that does not pose a significant safety or health hazard to employees in the immediate vicinity or have the potential to become an emergency within a short time. The period that constitutes a "short time" is difficult to define. Lab technicians can handle incidental spills because they are expected to be familiar with the hazards of the chemicals they routinely handle during an "average" workday. If the spill exceeds the scope of the lab technicians' experience, training, or willingness to respond, the workers must be able to determine that the spill cannot be dealt with internally.

Emergency assistance is typically provided by an outside agency. Spills requiring the involvement of individuals outside the lab are those exceeding the exposure one would expect during the normal course of work. Spills in this category are those that have truly become emergency situations in that lab technicians are overwhelmed beyond their level of training. Their response capability is compromised by the magnitude of the incident. OSHA elaborates that emergencies such as this involve:

- The need to evacuate employees in the area
- The need for response from outside the immediate release area
- A release that poses, or has potential to pose, conditions that are immediately dangerous to life and health
- A release that poses a serious threat of fire and explosion
- A release that requires immediate attention due to imminent danger
- A release that may cause high levels of exposure to toxic substances
- Uncertainty that the worker can handle the severity of the hazard with the personal protective equipment (PPE) and equipment that has been provided and the exposure limit could be exceeded easily
- A situation that is unclear or lacking data regarding important factors

Depending on the circumstances, what begins as an incidental spill could at some point escalate into a major emergency. Responding lab technicians must monitor changing conditions. Again, this clearly points to the need for lab-specific training.

Additional Reading

Csuros, M. *Environmental Sampling and Analysis for Technicians* (Boca Raton, Fla.: CRC Press, 1994).
Lohr, S. *Sampling: Design and Analysis* (Pacific Grove, Calif.: Brooks/Cole, 1998).

Summary

- Sampling and analysis constitute one of the major cornerstones of the EH&S profession. In performing sampling and analytic tasks, EH&S professionals:
  - use documented procedures and practices

- maintain control of samples that require chain of custody until the appropriate person has accepted responsibility for the samples and has signed the custody log
- follow sound laboratory procedures and any special instructions when analyzing a sample
- ensure all analytical equipment is calibrated and maintained according to QC procedures
- highlight any unusual or reportable monitoring results
- follow recommended good work practices
- The key elements involved in sampling include:
  - sampling plans
  - USEPA SOPs
  - representative samples
  - discrete and composite samples
  - background samples
  - blanks, splits
  - chain of custody
  - shipping requirements
- Laboratory analysis consists of two important subparts. Qualitative analysis is the determination of the identities, but not the concentrations, of the constituents present in a substance, a mixture of substances, or a solution. The process of determining the concentrations of the species present or the percent composition of a mixture is called quantitative analysis.
- Laboratories generally follow standard Quality Assurance and Quality Control (QA/QC) procedures.
- Advanced technologies, such as chromatography and spectroscopy, currently allow for more detailed and accurate analysis.
- Good laboratory practices should be integral to any lab operation.

New Word Review

Background samples—samples collected to establish the natural concentrations of constituents in the local environment prior to the introduction of potential contamination.

Chain of custody—the sequence and identity of entities in possession, ensuring sample integrity from collections to data reporting.

Chromatography—an analytical technique that involves passing a liquid or gaseous mixture of compounds across a solid matrix to take advantage of the different "affinities" of each compound for the solid matrix.

Composite sample—a combination of discrete samples collected at the same sampling point at different times.

Data validation—a detailed analysis of a data package to determine data quality by checking various parameters.

Detection limit—the minimum concentration of an analyte that can be measured above the instrument background noise.

Field blank—a sample of purified water, often provided by the analytical laboratory, which is taken out in the field for the duration of the sampling, where it is collected and processed in the same manner as the samples being taken. The purpose of the field blank is to ensure that no contamination has resulted from sample-handling activities in the field.

Mean—the total of the values of a set of observations divided by the number of observations.

Median—the value of the central item when the data are arrayed in size.

Method detection limit—the lowest concentration that can be detected by an instrument with correction for the effects of sample matrix and method-specific parameters such as sample preparation.

Mode—the observation that occurs with the greatest frequency and thus is the most "fashionable" value.

Practical quantification limit—to the practical concentration that can be reliably achieved within specified limits of precision and accuracy during routine laboratory operating conditions.

Qualitative analysis—the determination of identities of the constituents present in a substance, a mixture of substances, or a solution.

Quality assurance (QA)—activities that provide a formalized system for evaluating the technical adequacy of sample collection and laboratory analysis activities.

Quality control (QC)—the routine application of procedures for obtaining data that are accurate, representative, comparable, and complete.

Quantitative analysis—the process of determining the concentrations of the species present or the percent composition of a mixture.

Range—the difference between the values of the highest and lowest terms.

Representative sample—a sample containing all the constituents present in the media or waste being sampled.

Spectroscopy—an analytical technique that helps chemists identify which functional groups are present.

Spike recoveries—a measure of the accuracy of the methods used for analyses.

Split sample—a sample that is split into two parts in the field in order to obtain an independent analysis of the sub-samples.

Trip blank—a sample that is prepared prior to the sampling event in the actual container and is stored with the investigative samples throughout the sampling event.

Chapter Review Questions

15.1.   A _____ is a sample containing all the constituents present in the media or waste being sampled.

15.2.   The term _____ refers to a combination of discrete samples collected at the same sampling point at different times.

15.3.   _____ are used to determine if samples were contaminated during storage and/or transportation back to the laboratory.

15.4. The primary purpose of a split sample is to measure sample-handling _____.

15.5. A _____ must follow all sample shipments to record the times, locations, and names of all persons taking possession of samples.

15.6. They are defined as three times the standard deviation of replicate spiked analyses: _____

15.7. Each metal has a characteristic _____ pattern when heated.

15.8. Spike recoveries are routinely determined as part of any good laboratory _____.

15.9. UV spectrophotometers measure change in intensity per wavelength as _____.

15.10. Metals also have characteristic _____ patterns when heated to very high temperatures.

Notes

1. *Monitoring Water Quality: Intensive Stream Biosurvey* (Washington, D.C.: USEPA), 1–35, http:www.epa.gov/owow/monitoring/volunteer/stream/vms43.html (accessed August 18, 2000).

2. USEPA. *Test Methods for Evaluating Solid Waste: Physical/Chemical Methods,* 3rd ed. Publ. No. SW-846, Off. Solid Waste and Emergency Response (Washington, D.C., 1986); USEPA. *NEIC Policies and Procedures.* EPA-330/978(001)/-R (rev. 1982).

3. Based on material in R. J. Fessenden et al. *Organic Chemistry,* 6th ed. (Pacific Grove, Calif.: Brooks/Cole, 1998).

Answers for Chapter Review Questions

CHAPTER 2

2.1. element
2.2. proton
2.3. electron
2.4. electron
2.5. molecules
2.6. neutron
2.7. proton, neutron
2.8. proton
2.9. 32
2.10. 12
2.11. atomic weight increases
2.12. 8
2.13. 12.01 + 2(16.00) = 44.01 (rounded to 44)
2.14. see the following table:

Element	No. of Protons	No. of Electrons	Charge
Ag	47	47	0
K^+	19	18	+1
Br^-	35	36	−1
Ca^{+2}	20	18	+2

2.15. (a) physical, (b) chemical, (c) chemical, (d) physical, and (e) physical
2.16. c
2.17. b
2.18. d
2.19. b
2.20. b
2.21. b

CHAPTER 3

3.1. 3, 7, 2
3.2. different

3.3.   molecule
3.4.   attract
3.5.   polar bond
3.6.   oppositely
3.7.   symbol
3.8.   subscript
3.9.   nonpolar
3.10. nonpolar
3.11. polar
3.12. one
3.13. $2 \times 1.0 = 2.0, 1 \times 16.0; 2.0 + 16.0 = FW = 18.0$
3.14. $1 \times 1.0 = 1.0, 1 \times 14.0 = 14.0, 3 \times 16.0 = 48.0; 1.0 + 48.0 = FW = 63.0$
3.15. formula wt. of NaOH is $23.0 + 16.0 + 1.0 = 40.0$ (using the equation in figure 3.3, we have $100/40.0 = 2.5$ gram-moles of NaOH)
3.16. electron

CHAPTER 4

4.1.   a
4.2.   c
4.3.   b
4.4.   b
4.5.   c
4.6.   b
4.7.   c
4.8.   b
4.9.   cast iron
4.10. ore
4.11. ammonia
4.12. because poisonous lead salts leach out when lead dissolves
4.13. oxygen
4.14. the mineral is softer than 2 to 2.5 on the Moh's scale

CHAPTER 5

5.1.   10
5.2.   9
5.3.   1
5.4.   6
5.5.   7
5.6.   5
5.7.   3

5.8.   2
5.9.   8
5.10. 4
5.11. d
5.12. c
5.13. a
5.14. C_3H_8
5.15. aromatic
5.16. b
5.17. carboxyl group
5.18. –CHO
5.19. alcohol
5.20. five

CHAPTER 6

6.1.   a
6.2.   mass
6.3.   combination
6.4.   gas
6.5.   increases
6.6.   more, faster
6.7.   catalyst
6.8.   solids
6.9.   exothermic
6.10. decomposition
6.11. energy
6.12. more
6.13. rate
6.14. do not
6.15. faster
6.16. energy
6.17. increase
6.18. combination
6.19. combination
6.20. double replacement

CHAPTER 7

7.1.   form 174
7.2.   mixtures
7.3.   before
7.4.   acute

7.5. combustible or flammable
7.6. TWA
7.7. flash point
7.8. melting point
7.9. ppm
7.10. synergism
7.11. 12 ppm
7.12. 14 ppm

CHAPTER 8

8.1. Na
8.2. H_2SO_4
8.3. 7
8.4. $CaCO_3$
8.5. base
8.6. changes
8.7. solid, liquid, gas
8.8. element
8.9. compound
8.10. periodic
8.11. solvent, solute
8.12. an atom or group of atoms that carry a positive or negative electric charge as a result of having lost or gained one or more electrons
8.13. colloid
8.14. turbidity
8.15. toxicity
8.16. organic
8.17. 0 to 14
8.18. measure of water's ability to neutralize an acid
8.19. calcium and magnesium
8.20. base

CHAPTER 9

9.1. no
9.2. primary
9.3. secondary
9.4. sulfur dioxide
9.5. CFCs
9.6. photochemical
9.7. CAA
9.8. acid rain

9.9.   pollution
9.10. nitrogen, oxygen
9.11. aerosols
9.12. dust
9.13. carbon monoxide
9.14. lead
9.15. greenhouse effect
9.16. aerosols
9.17. old
9.18. automobile
9.19. natural
9.20. greenhouse effect

CHAPTER 10

10.1. petroleum-contaminated soil
10.2. soils
10.3. extracting it
10.4. coal tar
10.5. leaching
10.6. protons
10.7. amendments
10.8. color
10.9. drainage
10.10. vapor pressure
10.11. biodegradation
10.12. animal feedlots

CHAPTER 11

11.1. pesticide
11.2. pesticides
11.3. nitrogen
11.4. proteins
11.5. ammonia
11.6. alkaloids
11.7. ammonium nitrate
11.8. Paraquat
11.9. birds and reptiles
11.10. tricalcium phosphate
11.11. sulfur
11.12. sulfuric acid
11.13. sulfur oxides, sulfur trioxide

11.14. thiol/mercaptans
11.15. FIFRA
11.16. fumigant
11.17. barricade, absorb, neutralize

CHAPTER 12

12.1. acceptable risk
12.2. toxicology
12.3. type of organism exposed, amount of substance, rate of exposure by inhalation, injection, or skin contact
12.4. toxicological chemistry
12.5. dose
12.6. hyposensitivity
12.7. primary
12.8. selective toxicity
12.9. TLVs
12.10. target organs
12.11. dioxin
12.12. risk
12.13. risk assessment
12.14. epidemiology studies
12.15. TSCA (Toxic Substances Control Act)

CHAPTER 13

13.1. Roentgen
13.2. alpha particles
13.3. beta particles
13.4. gamma rays
13.5. alpha particle
13.6. radon
13.7. NORM
13.8. fission
13.9. fusion
13.10. half-life
13.11. mixed waste
13.12. radioactive isotope
13.13. see the following table:

% of radioactive source left	Mass of original sample (g)	Time
100%	750 g	0
50%	375 g	12.4 hr

25%	187 g	24.8 hr
12.5%	93.7 g	37.2 hr
6.25%	46.9 g	49.6 hr
3.125%	23.4 g	62.0 hr

13.14. see the following table (the time for one half-life is 7.2 seconds):

% of radioactive source left	Mass of original sample (g)	Time
100%	50.0 g	0
50%	25.0 g	
25%	12.5 g	14.4 sec

CHAPTER 14

14.1. hazardous waste
14.2. cradle-to-grave
14.3. reactivity
14.4. corrosivity
14.5. toxicity
14.6. silver
14.7. F list
14.8. physical treatment methods
14.9. sedimentation
14.10. oxidized

CHAPTER 15

15.1. representative sample
15.2. composite sample
15.3. trip blanks
15.4. variability
15.5. chain of custody form
15.6. method detection limit
15.7. absorption
15.8. QA/QC
15.9. absorbance
15.10. emission

About the Author

Frank R. Spellman is assistant professor of environmental health at Old Dominion University in Norfolk, Virginia. He has more than thirty-five years of experience in environmental science and engineering, in both the military and the civilian communities. A professional member of the American Society of Safety Engineers, the Water Environment Federation, and the Institute of Hazardous Materials Managers, Frank Spellman is a board-certified safety professional (CSP) and board-certified hazardous materials manager (CHMM). He has either authored or coauthored forty-seven texts on safety, occupational health, water/wastewater operations, environmental science, and concentrated animal feeding operations (CAFO).